I0044438

Greenhouse Gas: Science and Technology

Greenhouse Gas: Science and Technology

Edited by
Micah Taylor

Larsen & Keller
www.larsen-keller.com

Greenhouse Gas: Science and Technology
Edited by Micah Taylor
ISBN: 978-1-63549-141-8 (Hardback)

© 2017 Larsen & Keller

☰ Larsen & Keller

Published by Larsen and Keller Education,
5 Penn Plaza,
19th Floor,
New York, NY 10001, USA

Cataloging-in-Publication Data

Greenhouse gas : science and technology / edited by Micah Taylor.
     p. cm.
Includes bibliographical references and index.
ISBN 978-1-63549-141-8
1. Greenhouse gases. 2. Greenhouse gases--Environmental aspects. 3. Greenhouse gases--Technological innovations. 4. Greenhouse gases--Control--Technological innovations. I. Taylor, Micah.
TD885.5.G73 G74 2017
363.738 74--dc23

This book contains information obtained from authentic and highly regarded sources. All chapters are published with permission under the Creative Commons Attribution Share Alike License or equivalent. A wide variety of references are listed. Permissions and sources are indicated; for detailed attributions, please refer to the permissions page. Reasonable efforts have been made to publish reliable data and information, but the authors, editors and publisher cannot assume any responsibility for the vailidity of all materials or the consequences of their use.

Trademark Notice: All trademarks used herein are the property of their respective owners. The use of any trademark in this text does not vest in the author or publisher any trademark ownership rights in such trademarks, nor does the use of such trademarks imply any affiliation with or endorsement of this book by such owners.

The publisher's policy is to use permanent paper from mills that operate a sustainable forestry policy. Furthermore, the publisher ensures that the text paper and cover boards used have met acceptable environmental accreditation standards.

Printed and bound in the United States of America.

For more information regarding Larsen and Keller Education and its products, please visit the publisher's website www.larsen-keller.com

Table of Contents

Preface

This book is a valuable compilation of topics, ranging from the basic to the most complex theories and principles in the field of greenhouse gas and their effects on environment. It discusses in detail the problems caused by greenhouse gases and how to control the spread of the same. Greenhouse gases absorb and then emit radiations in a thermal infrared range and thus, causing greenhouse effect. The main types of greenhouse gases on Earth are nitrous oxide, carbon dioxide, methane, ozone and even water vapor. Also included in this text is a detailed explanation of the various concepts of greenhouse gas effect. Some of the diverse topics covered in it address the varied branches that fall under this subject. This textbook is a complete source of knowledge on the present status of this important field.

To facilitate a deeper understanding of the contents of this book a short introduction of every chapter is written below:

Chapter 1- Greenhouse gases were discovered in the past few decades and ever since, conscious effort has been made to reduce greenhouse gas emissions and the greenhouse effect. Current projections of greenhouse gas accumulation would lead to drastic changes in the earth's climate. This chapter is an overview of the subject matter incorporating all the major aspects of greenhouse gases.

Chapter 2- The emission of certain toxic gases as a by-product of human activity and their subsequent chemical reaction with particles in the atmosphere has led to those gases being classified as greenhouse gases. These gases radiate energy back to the earth's surface, causing global warming. This chapter lists some of the major greenhouse gases such as carbon dioxide, methane and nitrous oxide.

Chapter 3- Greenhouse gases and aerosols enter the earth's higher atmospheric levels and allow solar radiation to penetrate the earth's surface. These gases also re-emit solar energy onto the land and oceans, causing a rise in temperature levels. Some of the topics discussed in this chapter are radiative forcing and the greenhouse effect. This chapter elucidates the crucial theories and principles of greenhouse gas emissions.

Chapter 4- The most popular methods of reducing greenhouse gas emissions are through policy-making that targets corporations and industries that have the largest carbon footprint. Placing a cap on carbon expenditure is an effective way to control greenhouse gases. Carbon credit, carbon offset and carbon neutrality are some of the topics in this chapter. Green energy is best understood in confluence with the major topics listed in the following chapter.

Chapter 5- Greenhouse gases contribute greatly to the recent change in climate and there is an arising need to assess it in terms of global warming and climate change. National and international policies and protocols based on greenhouse gas emissions depend on these projections for policy-making and trade.

Chapter 6- Carbon accounting is a method for measuring greenhouse gas emissions. It has been taken up by corporations and firms that aim to reduce greenhouse gas emission as well as stimulate demand for carbon credit commodities. However, accurate measurement of greenhouse gases is still extremely difficult. This chapter discusses the methods of greenhouse gas measurement in a critical manner, providing key analysis to the subject matter.

I owe the completion of this book to the never-ending support of my family, who supported me throughout the project.

Editor

Introduction to Greenhouse Gas

Greenhouse gases were discovered in the past few decades and ever since, conscious effort has been made to reduce greenhouse gas emissions and the greenhouse effect. Current projections of greenhouse gas accumulation would lead to drastic changes in the earth's climate. This chapter is an overview of the subject matter incorporating all the major aspects of greenhouse gases.

A greenhouse gas (sometimes abbreviated GHG) is a gas in an atmosphere that absorbs and emits radiation within the thermal infrared range. This process is the fundamental cause of the greenhouse effect. The primary greenhouse gases in Earth's atmosphere are water vapor, carbon dioxide, methane, nitrous oxide, and ozone. Without greenhouse gases, the average temperature of Earth's surface would be about −18 °C (0 °F), rather than present average of 15 °C (59 °F). In the Solar System, the atmospheres of Venus, Mars and Titan also contain gases that cause a greenhouse effect.

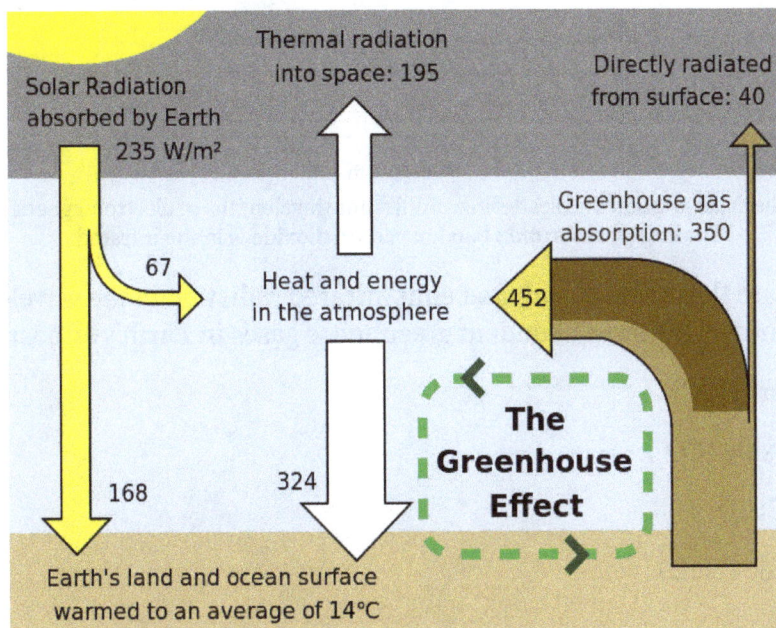

Greenhouse effect schematic showing energy flows between space, the atmosphere, and Earth's surface. Energy influx and emittance are expressed in watts per square meter (W/m^2).

Human activities since the beginning of the Industrial Revolution (taken as the year 1750) have produced a 40% increase in the atmospheric concentration of carbon dioxide, from 280 ppm in 1750 to 400 ppm in 2015. This increase has occurred despite the uptake of a large portion of the emissions by various natural "sinks" involved in the carbon cycle. Anthropogenic carbon dioxide (CO_2) emissions (i.e. emissions produced by human activities) come from combustion of carbon-based fuels, principally coal, oil, and natural gas, along with deforestation, soil erosion and animal agriculture.

It has been estimated that if greenhouse gas emissions continue at the present rate, Earth's surface temperature could exceed historical values as early as 2047, with potentially harmful effects on ecosystems, biodiversity and the livelihoods of people worldwide.

Gases in Earth's Atmosphere

Greenhouse Gases

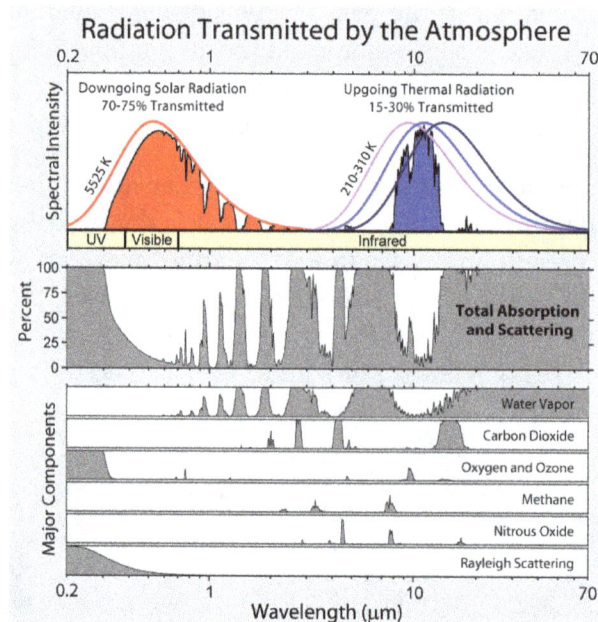

Atmospheric absorption and scattering at different wavelengths of electromagnetic waves.
The largest absorption band of carbon dioxide is in the infrared.

Greenhouse gases are those that absorb and emit infrared radiation in the wavelength range emitted by Earth. z In order, the most abundant greenhouse gases in Earth's atmosphere are:

- Water vapor (H_2O)

- Carbon dioxide (CO_2)

- Methane (CH_4)

- Nitrous oxide (N_2O)

- Ozone (O_3)

- Chlorofluorocarbons (CFCs)

Atmospheric concentrations of greenhouse gases are determined by the balance between sources (emissions of the gas from human activities and natural systems) and sinks (the removal of the gas from the atmosphere by conversion to a different chemical compound). The proportion of an emission remaining in the atmosphere after a specified time is the "airborne fraction" (AF). More precisely, the annual airborne fraction is the ratio of the atmospheric increase in a given year to that year's total emissions. Over the last 50 years (1956–2006) the airborne fraction for CO_2 has been increasing at 0.25 ± 0.21%/year.

Non-greenhouse Gases

The major atmospheric constituents, nitrogen (N2), oxygen (O2), and argon (Ar), are not greenhouse gases. This is because molecules containing two atoms of the same element such as N2 and O2 and monatomic molecules such as argon (Ar) have no net change in the distribution of their electrical charges when they vibrate and hence are almost totally unaffected by infrared radiation. Although molecules containing two atoms of different elements such as carbon monoxide (CO) or hydrogen chloride (HCl) absorb infrared radiation, these molecules are short-lived in the atmosphere owing to their reactivity and solubility. Therefore, they do not contribute significantly to the greenhouse effect and usually are omitted when discussing greenhouse gases.

Indirect Radiative Effects

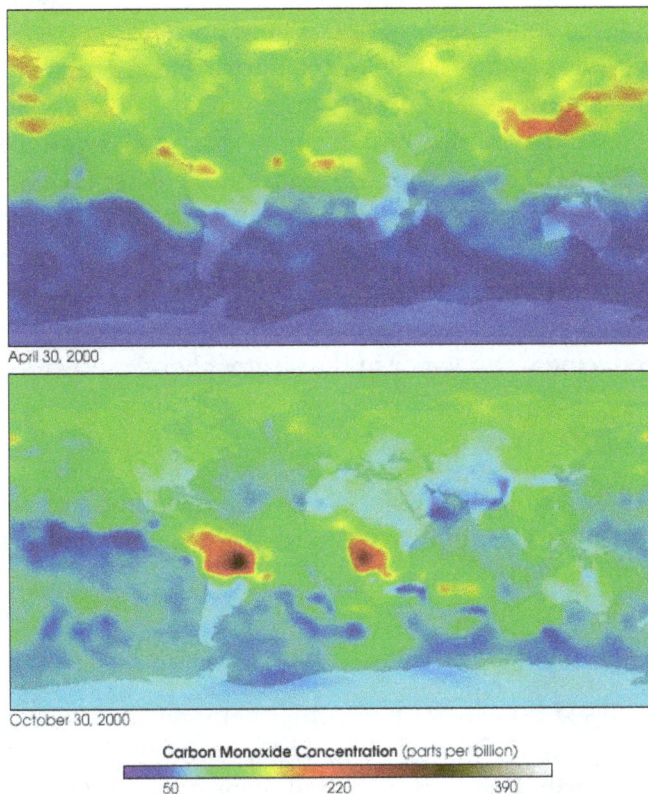

April 30, 2000

October 30, 2000

Carbon Monoxide Concentration (parts per billion)

50   220   390

The false colors in this image represent concentrations of carbon monoxide in the lower atmosphere, ranging from about 390 parts per billion (dark brown pixels), to 220 parts per billion (red pixels), to 50 parts per billion (blue pixels).

Some gases have indirect radiative effects (whether or not they are greenhouse gases themselves). This happens in two main ways. One way is that when they break down in the atmosphere they produce another greenhouse gas. For example, methane and carbon monoxide (CO) are oxidized to give carbon dioxide (and methane oxidation also produces water vapor; that will be considered below). Oxidation of CO to CO_2 directly produces an unambiguous increase in radiative forcing although the reason is subtle. The peak of the thermal IR emission from Earth's surface is very close to a strong vibrational absorption band of CO_2 (667 cm^{-1}). On the other hand, the single CO vibrational band only absorbs IR at much higher frequencies (2145 cm^{-1}), where the ~300 K thermal emission of the surface is at least a factor of ten lower. On the other hand, oxidation of methane to CO_2, which requires reactions with the OH radical, produces an instantaneous reduction, since CO_2 is a weaker greenhouse gas than methane; but it has a longer lifetime. As described below this is not the whole story, since the oxidations of CO and CH 4 are intertwined by both consuming OH radicals. In any case, the calculation of the total radiative effect needs to include both the direct and indirect forcing.

A second type of indirect effect happens when chemical reactions in the atmosphere involving these gases change the concentrations of greenhouse gases. For example, the destruction of non-methane volatile organic compounds (NMVOCs) in the atmosphere can produce ozone. The size of the indirect effect can depend strongly on where and when the gas is emitted.

Methane has a number of indirect effects in addition to forming CO_2. Firstly, the main chemical that destroys methane in the atmosphere is the hydroxyl radical (OH). Methane reacts with OH and so more methane means that the concentration of OH goes down. Effectively, methane increases its own atmospheric lifetime and therefore its overall radiative effect. The second effect is that the oxidation of methane can produce ozone. Thirdly, as well as making CO_2 the oxidation of methane produces water; this is a major source of water vapor in the stratosphere, which is otherwise very dry. CO and NMVOC also produce CO_2 when they are oxidized. They remove OH from the atmosphere and this leads to higher concentrations of methane. The surprising effect of this is that the global warming potential of CO is three times that of CO_2. The same process that converts NMVOC to carbon dioxide can also lead to the formation of tropospheric ozone. Halocarbons have an indirect effect because they destroy stratospheric ozone. Finally hydrogen can lead to ozone production and CH 4 increases as well as producing water vapor in the stratosphere.

Contribution of Clouds to Earth's Greenhouse Effect

The major non-gas contributor to Earth's greenhouse effect, clouds, also absorb and emit infrared radiation and thus have an effect on radiative properties of the greenhouse gases. Clouds are water droplets or ice crystals suspended in the atmosphere.

Impacts on the Overall Greenhouse Effect

Schmidt *et al.* (2010) analysed how individual components of the atmosphere contribute to the total greenhouse effect. They estimated that water vapor accounts for about 50% of Earth's greenhouse effect, with clouds contributing 25%, carbon dioxide 20%, and the minor greenhouse gases and aerosols accounting for the remaining 5%. In the study, the reference model atmosphere is for 1980 conditions.

The contribution of each gas to the greenhouse effect is affected by the characteristics of that gas, its abundance, and any indirect effects it may cause. For example, the direct radiative effect of a mass of methane is about 72 times stronger than the same mass of carbon dioxide over a 20-year time frame but it is present in much smaller concentrations so that its total direct radiative effect is smaller, in part due to its shorter atmospheric lifetime. On the other hand, in addition to its direct radiative impact, methane has a large, indirect radiative effect because it contributes to ozone formation. Shindell *et al.* (2005) argue that the contribution to climate change from methane is at least double previous estimates as a result of this effect.

When ranked by their direct contribution to the greenhouse effect, the most important are:

Compound	Formula	Concentration inatmosphere (ppm)	Contribution (%)
Water vapor and clouds	$H2O$	10–50,000[A]	36–72%
Carbon dioxide	CO_2	~400	9–26%
Methane	CH4	~1.8	4–9%
Ozone	O3	2–8[B]	3–7%

notes:
[A] Water vapor strongly varies locally
[B] The concentration in stratosphere. About 90% of the ozone in Earth's atmosphere is contained in the stratosphere.

In addition to the main greenhouse gases listed above, other greenhouse gases include sulfur hexafluoride, hydrofluorocarbons and perfluorocarbons. Some greenhouse gases are not often listed. For example, nitrogen trifluoride has a high global warming potential (GWP) but is only present in very small quantities.

Proportion of Direct Effects at a Given Moment

It is not possible to state that a certain gas causes an exact percentage of the greenhouse effect. This is because some of the gases absorb and emit radiation at the same frequencies as others, so that the total greenhouse effect is not simply the sum of the influence of each gas. The higher ends of the ranges quoted are for each gas alone; the lower ends account for overlaps with the other gases. In addition, some gases such as methane are known to have large indirect effects that are still being quantified.

Atmospheric Lifetime

Aside from water vapor, which has a residence time of about nine days, major greenhouse gases are well mixed and take many years to leave the atmosphere. Although it is not easy to know with precision how long it takes greenhouse gases to leave the atmosphere, there are estimates for the principal greenhouse gases. Jacob (1999) defines the lifetime τ of an atmospheric species X in a one-box model as the average time that a molecule of X remains in the box. Mathematically τ can be defined as the ratio of the mass m (in kg) of X in the box to its removal rate, which is the sum of the flow of X out of the box F_{out}), chemical loss of X (L), and deposition of X (D) (all in kg/s):

$\tau = \dfrac{m}{F_{out} + L + D}$. If one stopped pouring any of this gas into the box, then after a time τ τ, its concentration would be about halved.

The atmospheric lifetime of a species therefore measures the time required to restore equilibrium following a sudden increase or decrease in its concentration in the atmosphere. Individual atoms or molecules may be lost or deposited to sinks such as the soil, the oceans and other waters, or vegetation and other biological systems, reducing the excess to background concentrations. The average time taken to achieve this is the mean lifetime.

Carbon dioxide has a variable atmospheric lifetime, and cannot be specified precisely. The atmospheric lifetime of CO_2 is estimated of the order of 30–95 years. This figure accounts for CO_2 molecules being removed from the atmosphere by mixing into the ocean, photosynthesis, and other processes. However, this excludes the balancing fluxes of CO_2 into the atmosphere from the geological reservoirs, which have slower characteristic rates. Although more than half of the CO_2 emitted is removed from the atmosphere within a century, some fraction (about 20%) of emitted CO_2 remains in the atmosphere for many thousands of years.   Similar issues apply to other greenhouse gases, many of which have longer mean lifetimes than CO_2. E.g., N_2O has a mean atmospheric lifetime of 114 years.

Radiative Forcing

Earth absorbs some of the radiant energy received from the sun, reflects some of it as light and reflects or radiates the rest back to space as heat. Earth's surface temperature depends on this balance between incoming and outgoing energy. If this energy balance is shifted, Earth's surface could become warmer or cooler, leading to a variety of changes in global climate.

A number of natural and man-made mechanisms can affect the global energy balance and force changes in Earth's climate. Greenhouse gases are one such mechanism. Greenhouse gases in the atmosphere absorb and re-emit some of the outgoing energy radiated from Earth's surface, causing that heat to be retained in the lower atmosphere. As explained above, some greenhouse gases remain in the atmosphere for decades or even centuries, and therefore can affect Earth's energy balance over a long time period. Factors that influence Earth's energy balance can be quantified in terms of "radiative climate forcing." Positive radiative forcing indicates warming (for example, by increasing incoming energy or decreasing the amount of energy that escapes to space), whereas negative forcing is associated with cooling.

Global Warming Potential

The global warming potential (GWP) depends on both the efficiency of the molecule as a greenhouse gas and its atmospheric lifetime. GWP is measured relative to the same mass of CO_2 and evaluated for a specific timescale. Thus, if a gas has a high (positive) radiative forcing but also a short lifetime, it will have a large GWP on a 20-year scale but a small one on a 100-year scale. Conversely, if a molecule has a longer atmospheric lifetime than CO_2 its GWP will increase with the timescale considered. Carbon dioxide is defined to have a GWP of 1 over all time periods.

 Examples of the atmospheric lifetime and GWP relative to CO_2 for several greenhouse gases are given in the following table:

Atmospheric lifetime and GWP relative to CO_2 at different time horizon for various greenhouse gases.					
Gas name	**Chemical formula**	**Lifetime (years)**	**Global warming potential (GWP) for given time horizon**		
			20-yr	**100-yr**	**500-yr**
Carbon dioxide	CO_2	30–95	1	1	1
Methane	CH4	12	72	25	7.6

Nitrous oxide	N2O	114	289	298	153
CFC-12	CCl2F2	100	11 000	10 900	5 200
HCFC-22	CHClF2	12	5 160	1 810	549
Tetrafluoromethane	CF4	50 000	5 210	7 390	11 200
Hexafluoroethane	C2F6	10 000	8 630	12 200	18 200
Sulfur hexafluoride	SF6	3 200	16 300	22 800	32 600
Nitrogen trifluoride	NF3	740	12 300	17 200	20 700

The use of CFC-12 (except some essential uses) has been phased out due to its ozone depleting properties. The phasing-out of less active HCFC-compounds will be completed in 2030.

Carbon dioxide in Earth's atmosphere if *half* of global-warming emissions are *not* absorbed.
(NASA simulation; 9 November 2015)

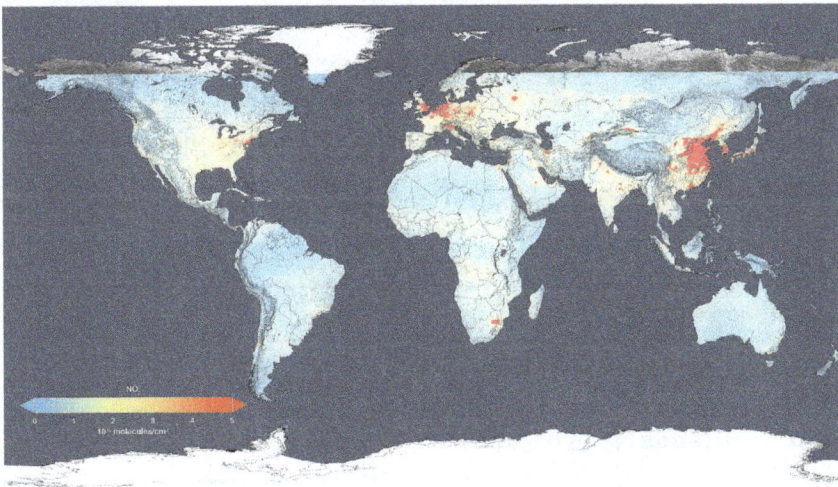

Nitrogen dioxide 2014 – global air quality levels
(released 14 December 2015).

Natural and Anthropogenic Sources

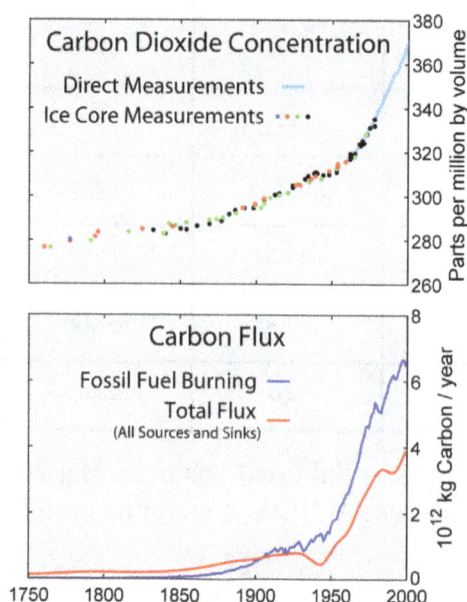

Top: Increasing atmospheric carbon dioxide levels as measured in the atmosphere and reflected in ice cores. Bottom: The amount of net carbon increase in the atmosphere, compared to carbon emissions from burning fossil fuel.

This diagram shows a simplified representation of the contemporary global carbon cycle. Changes are measured in gigatons of carbon per year (GtC/y). Canadell *et al.* (2007) estimated the growth rate of global average atmospheric CO_2 for 2000–2006 as 1.93 parts-per-million per year (4.1 petagrams of carbon per year).

Aside from purely human-produced synthetic halocarbons, most greenhouse gases have both natural and human-caused sources. During the pre-industrial Holocene, concentrations of existing gases were roughly constant. In the industrial era, human activities have added greenhouse gases to the atmosphere, mainly through the burning of fossil fuels and clearing of forests.

The 2007 Fourth Assessment Report compiled by the IPCC (AR4) noted that "changes in atmospheric concentrations of greenhouse gases and aerosols, land cover and solar radiation alter the energy balance of the climate system", and concluded that "increases in anthropogenic greenhouse gas concentrations is very likely to have caused most of the increases in global average temperatures since the mid-20th century". In AR4, "most of" is defined as more than 50%.

Abbreviations used in the two tables below: ppm = parts-per-million; ppb = parts-per-billion; ppt = parts-per-trillion; W/m² = watts per square metre

Current greenhouse gas concentrations					
Gas	Pre-1750 tropospheric concentration	Recent tropospheric concentration	Absolute increase since 1750	Percentage increase since 1750	Increased radiative forcing (W/m²)
Carbon dioxide (CO_2)	280 ppm	395.4 ppm	115.4 ppm	41.2%	1.88

Methane (CH4)	700 ppb	1893 ppb /1762 ppb	1193 ppb /1062 ppb	170.4% /151.7%	0.49
Nitrous oxide (N2O)	270 ppb	326 ppb /324 ppb	56 ppb /54 ppb	20.7% /20.0%	0.17
Tropospheric ozone (O3)	237 ppb	337 ppb	100 ppb	42%	0.4

Relevant to radiative forcing and/or ozone depletion; all of the following have no natural sources and hence zero amounts pre-industrial		
Gas	**Recent tropospheric concentration**	**Increased radiative forcing (W/m²)**
CFC-11 (trichlorofluoromethane)(CCl3F)	236 ppt /234 ppt	0.061
CFC-12 (CCl2F2)	527 ppt /527 ppt	0.169
CFC-113 (Cl2FC-CClF2)	74 ppt /74 ppt	0.022
HCFC-22 (CHClF2)	231 ppt /210 ppt	0.046
HCFC-141b (CH3CCl2F)	24 ppt /21 ppt	0.0036
HCFC-142b (CH3CClF2)	23 ppt /21 ppt	0.0042
Halon 1211 (CBrClF2)	4.1 ppt /4.0 ppt	0.0012
Halon 1301 (CBrClF3)	3.3 ppt /3.3 ppt	0.001
HFC-134a (CH2FCF3)	75 ppt /64 ppt	0.0108
Carbon tetrachloride (CCl4)	85 ppt /83 ppt	0.0143
Sulfur hexafluoride (SF6)	7.79 ppt /7.39 ppt	0.0043
Other halocarbons	Varies bysubstance	collectively0.02
Halocarbons in total		0.3574

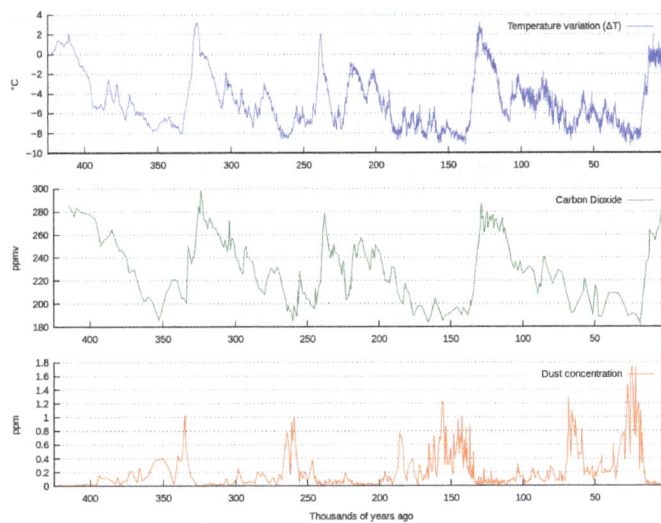

400,000 years of ice core data

Ice cores provide evidence for greenhouse gas concentration variations over the past 800,000 years (see the following section). Both CO$_2$ and CH4 vary between glacial and interglacial phases, and concentrations of these gases correlate strongly with temperature. Direct data does not exist for periods earlier than those represented in the ice core record, a record that indicates CO_2 mole fractions stayed within a range of 180 ppm to 280 ppm throughout the last 800,000 years, until the increase of the last 250 years. However, various proxies and modeling suggests larger variations in past epochs; 500 million years ago CO_2 levels were likely 10 times higher than now. Indeed, higher CO_2 concentrations are thought to have prevailed throughout most of the Phanerozoic eon, with concentrations four to six times current concentrations during the Mesozoic era, and ten to fifteen times current concentrations during the early Palaeozoic era until the middle of the Devonian period, about 400 Ma. The spread of land plants is thought to have reduced CO_2 concentrations during the late Devonian, and plant activities as both sources and sinks of CO_2 have since been important in providing stabilising feedbacks. Earlier still, a 200-million year period of intermittent, widespread glaciation extending close to the equator (Snowball Earth) appears to have been ended suddenly, about 550 Ma, by a colossal volcanic outgassing that raised the CO_2 concentration of the atmosphere abruptly to 12%, about 350 times modern levels, causing extreme greenhouse conditions and carbonate deposition as limestone at the rate of about 1 mm per day. This episode marked the close of the Precambrian eon, and was succeeded by the generally warmer conditions of the Phanerozoic, during which multicellular animal and plant life evolved. No volcanic carbon dioxide emission of comparable scale has occurred since. In the modern era, emissions to the atmosphere from volcanoes are only about 1% of emissions from human sources.

Ice Cores

Measurements from Antarctic ice cores show that before industrial emissions started atmospheric CO_2 mole fractions were about 280 parts per million (ppm), and stayed between 260 and 280 during the preceding ten thousand years. Carbon dioxide mole fractions in the atmosphere have gone up by approximately 35 percent since the 1900s, rising from 280 parts per million by volume to 387 parts per million in 2009. One study using evidence from stomata of fossilized leaves suggests greater variability, with carbon dioxide mole fractions above 300 ppm during the period seven to ten thousand years ago, though others have argued that these findings more likely reflect calibration or contamination problems rather than actual CO_2 variability. Because of the way air is trapped in ice (pores in the ice close off slowly to form bubbles deep within the firn) and the time period represented in each ice sample analyzed, these figures represent averages of atmospheric concentrations of up to a few centuries rather than annual or decadal levels.

Changes Since the Industrial Revolution

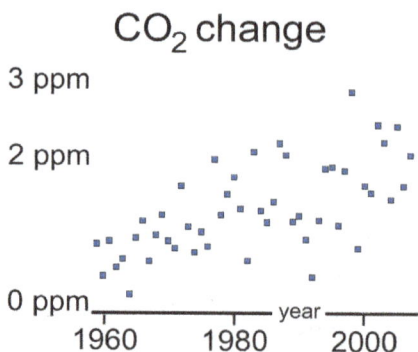

Recent year-to-year increase of atmospheric CO2.

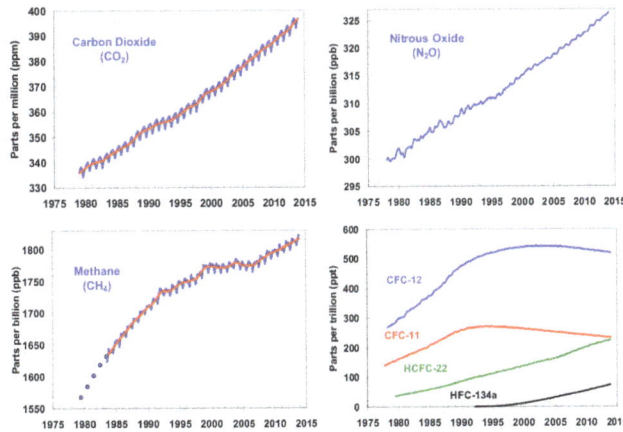

Major greenhouse gas trends.

Since the beginning of the Industrial Revolution, the concentrations of most of the greenhouse gases have increased. For example, the mole fraction of carbon dioxide has increased from 280 ppm by about 36% to 380 ppm, or 100 ppm over modern pre-industrial levels. The first 50 ppm increase took place in about 200 years, from the start of the Industrial Revolution to around 1973. however the next 50 ppm increase took place in about 33 years, from 1973 to 2006.

Recent data also shows that the concentration is increasing at a higher rate. In the 1960s, the average annual increase was only 37% of what it was in 2000 through 2007.

Today, the stock of carbon in the atmosphere increases by more than 3 million tonnes per annum (0.04%) compared with the existing stock. This increase is the result of human activities by burning fossil fuels, deforestation and forest degradation in tropical and boreal regions.

The other greenhouse gases produced from human activity show similar increases in both amount and rate of increase. Many observations are available online in a variety of Atmospheric Chemistry Observational Databases.

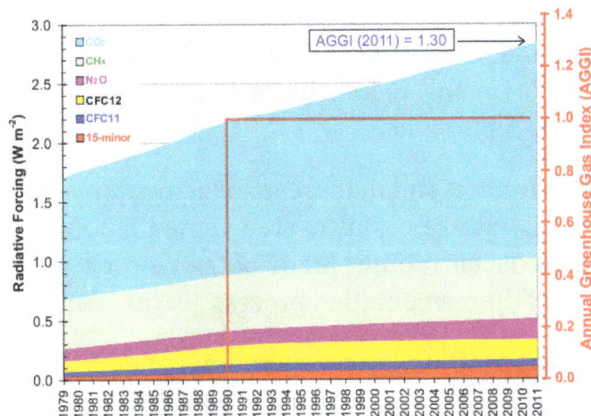

This graph shows changes in the annual greenhouse gas index (AGGI) between 1979 and 2011. The AGGI measures the levels of greenhouse gases in the atmosphere based on their ability to cause changes in Earth's climate.

This bar graph shows global greenhouse gas emissions by sector from 1990 to 2005, measured in carbon dioxide equivalents.

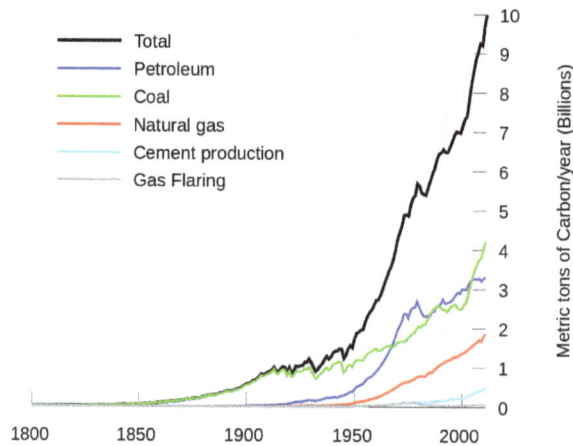

Modern global CO2 emissions from the burning of fossil fuels.

Since about 1750 human activity has increased the concentration of carbon dioxide and other greenhouse gases. Measured atmospheric concentrations of carbon dioxide are currently 100 ppm higher than pre-industrial levels. Natural sources of carbon dioxide are more than 20 times greater than sources due to human activity, but over periods longer than a few years natural sources are closely balanced by natural sinks, mainly photosynthesis of carbon compounds by plants and marine plankton. As a result of this balance, the atmospheric mole fraction of carbon dioxide remained between 260 and 280 parts per million for the 10,000 years between the end of the last glacial maximum and the start of the industrial era.

It is likely that anthropogenic (i.e., human-induced) warming, such as that due to elevated greenhouse gas levels, has had a discernible influence on many physical and biological systems. Future warming is projected to have a range of impacts, including sea level rise, increased frequencies and severities of some extreme weather events, loss of biodiversity, and regional changes in agricultural productivity.

The main sources of greenhouse gases due to human activity are:

- burning of fossil fuels and deforestation leading to higher carbon dioxide concentrations in the air. Land use change (mainly deforestation in the tropics) account for up to one third of total anthropogenic CO_2 emissions.

- livestock enteric fermentation and manure management, paddy rice farming, land use and wetland changes, pipeline losses, and covered vented landfill emissions leading to higher methane atmospheric concentrations. Many of the newer style fully vented septic systems that enhance and target the fermentation process also are sources of atmospheric methane.

- use of chlorofluorocarbons (CFCs) in refrigeration systems, and use of CFCs and halons in fire suppression systems and manufacturing processes.

- agricultural activities, including the use of fertilizers, that lead to higher nitrous oxide (N 2O) concentrations.

The seven sources of CO_2 from fossil fuel combustion are (with percentage contributions for 2000–2004):

Seven main fossil fuelcombustion sources	Contribution(%)
Liquid fuels (e.g., gasoline, fuel oil)	36%
Solid fuels (e.g., coal)	35%
Gaseous fuels (e.g., natural gas)	20%
Cement production	3 %
Flaring gas industrially and at wells	< 1%
Non-fuel hydrocarbons	< 1%
"International bunker fuels" of transportnot included in national inventories	4 %

Carbon dioxide, methane, nitrous oxide (N2O) and three groups of fluorinated gases (sulfur hexa-fluoride (SF6), hydrofluorocarbons (HFCs), and perfluorocarbons (PFCs)) are the major anthropogenic greenhouse gases, and are regulated under the Kyoto Protocol international treaty, which came into force in 2005. Emissions limitations specified in the Kyoto Protocol expired in 2012. The Cancún agreement, agreed in 2010, includes voluntary pledges made by 76 countries to control emissions. At the time of the agreement, these 76 countries were collectively responsible for 85% of annual global emissions.

Although CFCs are greenhouse gases, they are regulated by the Montreal Protocol, which was motivated by CFCs' contribution to ozone depletion rather than by their contribution to global warming. Note that ozone depletion has only a minor role in greenhouse warming though the two processes often are confused in the media.

Sectors

Tourism

According to UNEP global tourism is closely linked to climate change. Tourism is a significant contributor to the increasing concentrations of greenhouse gases in the atmosphere. Tourism accounts for about 50% of traffic movements. Rapidly expanding air traffic contributes about 2.5% of the production of CO_2. The number of international travelers is expected to increase from 594 million in 1996 to 1.6 billion by 2020, adding greatly to the problem unless steps are taken to reduce emissions.

Road Haulage

The road haulage industry plays a part in production of CO_2, contributing around 20% of the UK's total carbon emissions a year, with only the energy industry having a larger impact at around 39%. Average carbon emissions within the haulage industry are falling—in the thirty-year period from 1977–2007, the carbon emissions associated with a 200-mile journey fell by 21 percent; NOx emissions are also down 87 percent, whereas journey times have fallen by around a third. Due to their size, HGVs often receive criticism regarding their CO2 emissions; however, rapid development in engine technology and fuel management is having a largely positive effect.

Role of Water Vapor

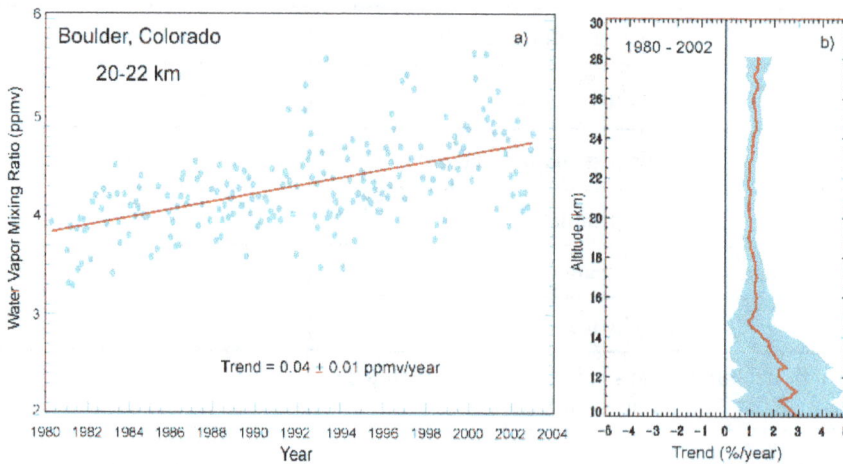

Increasing water vapor in the stratosphere at Boulder, Colorado.

Water vapor accounts for the largest percentage of the greenhouse effect, between 36% and 66% for clear sky conditions and between 66% and 85% when including clouds. Water vapor concentrations fluctuate regionally, but human activity does not significantly affect water vapor concentrations except at local scales, such as near irrigated fields. The atmospheric concentration of vapor is highly variable and depends largely on temperature, from less than 0.01% in extremely cold regions up to 3% by mass in saturated air at about 32 °C.

The average residence time of a water molecule in the atmosphere is only about nine days, compared to years or centuries for other greenhouse gases such as CH4 and CO_2. Thus, water vapor responds to and amplifies effects of the other greenhouse gases. The Clausius–Clapeyron relation establishes that more water vapor will be present per unit volume at elevated temperatures. This and other basic principles indicate that warming associated with increased concentrations of the other greenhouse gases also will increase the concentration of water vapor (assuming that the relative humidity remains approximately constant; modeling and observational studies find that this is indeed so). Because water vapor is a greenhouse gas, this results in further warming and so is a "positive feedback" that amplifies the original warming. Eventually other earth processes offset these positive feedbacks, stabilizing the global temperature at a new equilibrium and preventing the loss of Earth's water through a Venus-like runaway greenhouse effect.

Direct Greenhouse Gas Emissions

Between the period 1970 to 2004, GHG emissions (measured in CO_2-equivalent) increased at an average rate of 1.6% per year, with CO_2 emissions from the use of fossil fuels growing at a rate of 1.9% per year. Total anthropogenic emissions at the end of 2009 were estimated at 49.5 gigatonnes CO_2-equivalent. These emissions include CO_2 from fossil fuel use and from land use, as well as emissions of methane, nitrous oxide and other GHGs covered by the Kyoto Protocol.

At present, the primary source of CO_2 emissions is the burning of coal, natural gas, and petroleum for electricity and heat.

Regional and National Attribution of Emissions

Annual Greenhouse Gas Emissions by Sector

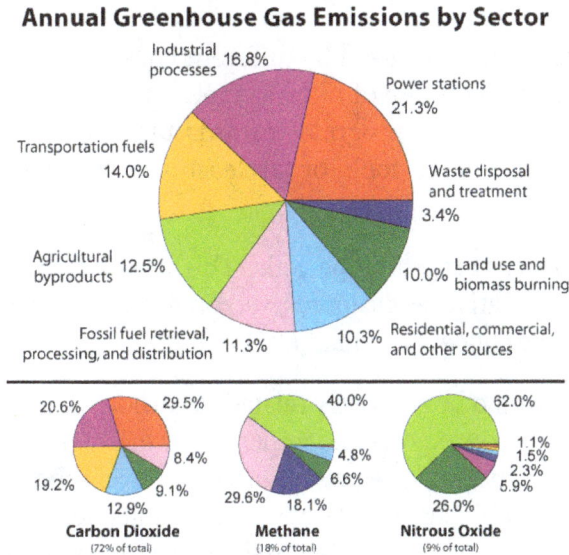

This figure shows the relative fraction of anthropogenic greenhouse gases coming from each of eight categories of sources, as estimated by the Emission Database for Global Atmospheric Research version 3.2, fast track 2000 project . These values are intended to provide a snapshot of global annual greenhouse gas emissions in the year 2000. The top panel shows the sum over all anthropogenic greenhouse gases, weighted by their global warming potential over the next 100 years. This consists of 72% carbon dioxide, 18% methane, 8% nitrous oxide and 1% other gases. Lower panels show the comparable information for each of these three primary greenhouse gases, with the same coloring of sectors as used in the top chart. Segments with less than 1% fraction are not labeled.

There are several different ways of measuring GHG emissions for tables of national emissions data. Some variables that have been reported include:

- Definition of measurement boundaries: Emissions can be attributed geographically, to the area where they were emitted (the territory principle) or by the activity principle to the territory produced the emissions. These two principles result in different totals when measuring, for example, electricity importation from one country to another, or emissions at an international airport.

- Time horizon of different GHGs: Contribution of a given GHG is reported as a CO_2 equivalent. The calculation to determine this takes into account how long that gas remains in the atmosphere. This is not always known accurately and calculations must be regularly updated to reflect new information.

- What sectors are included in the calculation (e.g., energy industries, industrial processes, agriculture etc.): There is often a conflict between transparency and availability of data.

- The measurement protocol itself: This may be via direct measurement or estimation. The four main methods are the emission factor-based method, mass balance method, predictive emissions monitoring systems, and continuous emissions monitoring systems. These methods differ in accuracy, cost, and usability.

These different measures are sometimes used by different countries to assert various policy/ethical positions on climate change. This use of different measures leads to a lack of comparability, which is problematic when monitoring progress towards targets. There are arguments for the adoption of a common measurement tool, or at least the development of communication between different tools.

Emissions may be measured over long time periods. This measurement type is called historical or cumulative emissions. Cumulative emissions give some indication of who is responsible for the build-up in the atmospheric concentration of GHGs.

The national accounts balance would be positively related to carbon emissions. The national accounts balance shows the difference between exports and imports. For many richer nations, such as the United States, the accounts balance is negative because more goods are imported than they are exported. This is mostly due to the fact that it is cheaper to produce goods outside of developed countries, leading the economies of developed countries to become increasingly dependent on services and not goods. We believed that a positive accounts balance would means that more production was occurring in a coun-try, so more factories working would increase carbon emission levels.

Emissions may also be measured across shorter time periods. Emissions changes may, for example, be measured against a base year of 1990. 1990 was used in the United Nations Framework Convention on Climate Change (UNFCCC) as the base year for emissions, and is also used in the Kyoto Protocol (some gases are also measured from the year 1995). A country's emissions may also be reported as a proportion of global emissions for a particular year.

Another measurement is of per capita emissions. This divides a country's total annual emissions by its mid-year population. Per capita emissions may be based on historical or annual emissions.

Land-use Change

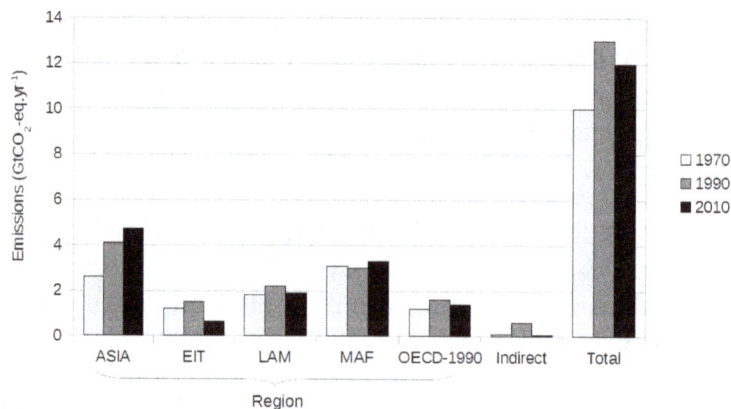

Greenhouse gas emissions from agriculture, forestry and other land use, 1970–2010.

Land-use change, e.g., the clearing of forests for agricultural use, can affect the concentration of GHGs in the atmosphere by altering how much carbon flows out of the atmosphere into carbon sinks. Accounting for land-use change can be understood as an attempt to measure "net" emissions, i.e., gross emissions from all GHG sources minus the removal of emissions from the atmosphere by carbon sinks.

There are substantial uncertainties in the measurement of net carbon emissions. Additionally, there is controversy over how carbon sinks should be allocated between different regions and over time. For instance, concentrating on more recent changes in carbon sinks is likely to favour those regions that have deforested earlier, e.g., Europe.

Greenhouse Gas Intensity

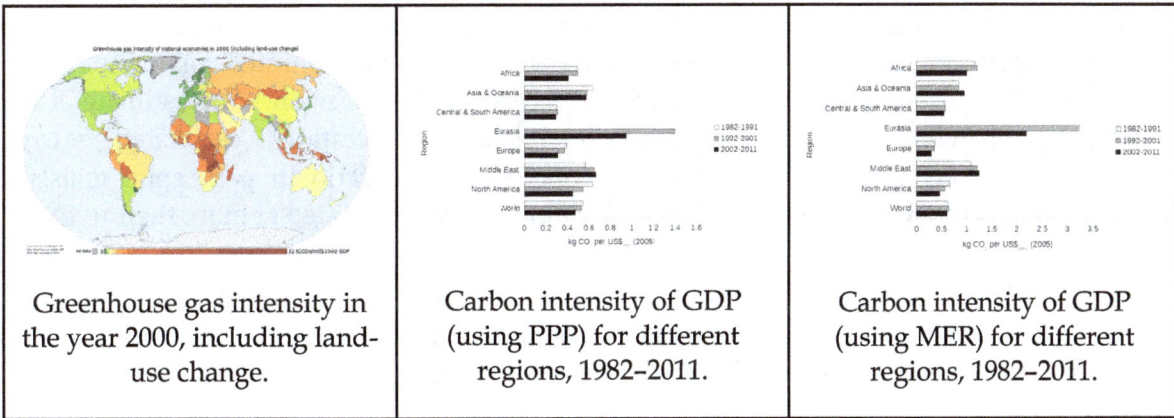

| Greenhouse gas intensity in the year 2000, including land-use change. | Carbon intensity of GDP (using PPP) for different regions, 1982–2011. | Carbon intensity of GDP (using MER) for different regions, 1982–2011. |

Greenhouse gas intensity is a ratio between greenhouse gas emissions and another metric, e.g., gross domestic product (GDP) or energy use. The terms "carbon intensity" and "emissions intensity" are also sometimes used. GHG intensities may be calculated using market exchange rates (MER) or purchasing power parity (PPP). Calculations based on MER show large differences in intensities between developed and developing countries, whereas calculations based on PPP show smaller differences.

Cumulative and Historical Emissions

Cumulative energy-related CO_2 emissions between the years 1850–2005 grouped into low-income, middle-income, high-income, the EU-15, and the OECD countries.

Cumulative energy-related CO_2 emissions between the years 1850–2005 for individual countries.

Map of cumulative per capita anthropogenic atmospheric CO_2 emissions by country. Cumulative emissions include land use change, and are measured between the years 1950 and 2000.

Regional trends in annual CO_2 emissions from fuel combustion between 1971 and 2009.

Regional trends in annual per capita CO_2 emissions from fuel combustion between 1971 and 2009.

Cumulative anthropogenic (i.e., human-emitted) emissions of CO_2 from fossil fuel use are a major cause of global warming, and give some indication of which countries have contributed most to human-induced climate change.

Top-5 historic CO_2 contributors by region over the years 1800 to 1988 (in %)		
Region	**Industrial CO_2**	**Total CO_2**
OECD North America	33.2	29.7
OECD Europe	26.1	16.6
Former USSR	14.1	12.5
China	5.5	6.0
Eastern Europe	5.5	4.8

Overall, developed countries accounted for 83.8% of industrial CO_2 emissions over this time period, and 67.8% of total CO_2 emissions. Developing countries accounted for industrial CO_2 emissions of 16.2% over this time period, and 32.2% of total CO_2 emissions. The estimate of total CO_2 emissions includes biotic carbon emissions, mainly from deforestation. calculated per capita cumulative emissions based on then-current population. The ratio in per capita emissions between industrialized countries and developing countries was estimated at more than 10 to 1.

Including biotic emissions brings about the same controversy mentioned earlier regarding carbon sinks and land-use change. The actual calculation of net emissions is very complex, and is affected by how carbon sinks are allocated between regions and the dynamics of the climate system.

Non-OECD countries accounted for 42% of cumulative energy-related CO_2 emissions between 1890–2007. Over this time period, the US accounted for 28% of emissions; the EU, 23%; Russia, 11%; China, 9%; other OECD countries, 5%; Japan, 4%; India, 3%; and the rest of the world, 18%.

Changes a Particular Base Year

Between 1970–2004, global growth in annual CO_2 emissions was driven by North America, Asia, and the Middle East. The sharp acceleration in CO_2 emissions since 2000 to more than a 3% increase per year (more than 2 ppm per year) from 1.1% per year during the 1990s is attributable to the lapse of formerly declining trends in carbon intensity of both developing and developed nations. China was responsible for most of global growth in emissions during this period. Localised plummeting emissions associated with the collapse of the Soviet Union have been followed by slow emissions growth in this region due to more efficient energy use, made necessary by the increasing proportion of it that is exported. In comparison, methane has not increased appreciably, and N 2O by 0.25% y^{-1}.

Using different base years for measuring emissions has an effect on estimates of national contributions to global warming. This can be calculated by dividing a country's highest contribution to global warming starting from a particular base year, by that country's minimum contribution to global warming starting from a particular base year. Choosing between different base years of 1750, 1900, 1950, and 1990 has a significant effect for most countries. Within the G8 group of countries, it is most significant for the UK, France and Germany. These countries have a long history of CO_2 emissions.

Annual Emissions

Annual per capita emissions in the industrialized countries are typically as much as ten times the average in developing countries. Due to China's fast economic development, its annual per capita emissions are quickly approaching the levels of those in the Annex I group of the Kyoto Protocol (i.e., the developed countries excluding the USA). Other countries with fast growing emissions are South Korea, Iran, and Australia (which apart from the oil rich Persian Gulf states, now has the highest percapita emission rate in the world). On the other hand, annual per capita emissions of the EU-15 and the USA are gradually decreasing over time. Emissions in Russia and Ukraine have decreased fastest since 1990 due to economic restructuring in these countries.

Energy statistics for fast growing economies are less accurate than those for the industrialized countries. For China's annual emissions in 2008, the Netherlands Environmental Assessment Agency estimated an uncertainty range of about 10%.

The GHG footprint, or greenhouse gas footprint, refers to the amount of GHG that are emitted during the creation of products or services. It is more comprehensive than the commonly used carbon footprint, which measures only carbon dioxide, one of many greenhouse gases.

2015 was the first year to see both total global economic growth and a reduction of carbon emissions.

Top Emitter Countries

The top 40 countries emitting all greenhouse gases, showing both that derived from all sources including land clearance and forestry and also the CO2 component excluding those sources. Per capita figures are included. Data taken from World Resources Institute, Washington. Note that Indonesia and Brazil show very much higher than on graphs simply showing fossil fuel use.

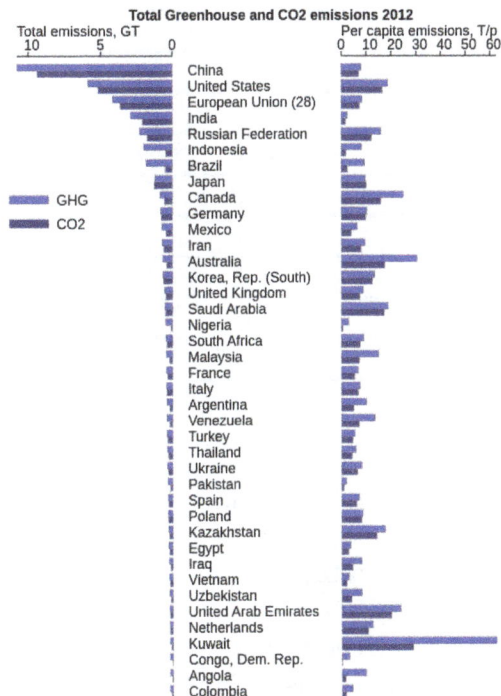

Total Greenhouse and CO2 emissions 2012

Annual

In 2009, the annual top ten emitting countries accounted for about two-thirds of the world's annual energy-related CO_2 emissions.

Top-10 annual energy-related CO_2 emitters for the year 2009		
Country	**% of global total annual emissions**	**Tonnes of GHG per capita**
People's Rep. of China	23.6	5.13
United States	17.9	16.9
India	5.5	1.37
Russian Federation	5.3	10.8
Japan	3.8	8.6
Germany	2.6	9.2
Islamic Rep. of Iran	1.8	7.3
Canada	1.8	15.4
South Korea	1.8	10.6
United Kingdom	1.6	7.5

Cumulative

Top-10 cumulative energy-related CO_2 emitters between 1850–2008		
Country	**% of world total**	**Metric tonnes CO_2 per person**
United States	28.5	1,132.7
China	9.36	85.4
Russian Federation	7.95	677.2
Germany	6.78	998.9
United Kingdom	5.73	1,127.8
Japan	3.88	367
France	2.73	514.9
India	2.52	26.7
Canada	2.17	789.2
Ukraine	2.13	556.4

Embedded Emissions

One way of attributing greenhouse gas (GHG) emissions is to measure the embedded emissions (also referred to as "embodied emissions") of goods that are being consumed. Emissions are usually measured according to production, rather than consumption. For example, in the main international treaty on climate change (the UNFCCC), countries report on emissions produced within their borders, e.g., the emissions produced from burning fossil fuels. Under a production-based accounting of emissions, embedded emissions on imported goods are attributed to the exporting, rather than the importing, country. Under a consumption-based accounting of emissions, embedded emissions on imported goods are attributed to the importing country, rather than the exporting, country.

Davis and Caldeira (2010) found that a substantial proportion of CO_2 emissions are traded internationally. The net effect of trade was to export emissions from China and other emerging markets to consumers in the US, Japan, and Western Europe. Based on annual emissions data from the year 2004, and on a per-capita consumption basis, the top-5 emitting countries were found to be (in tCO_2 per person, per year): Luxembourg (34.7), the US (22.0), Singapore (20.2), Australia (16.7), and Canada (16.6). Carbon Trust research revealed that approximately 25% of all CO_2 emissions from human activities 'flow' (i.e. are imported or exported) from one country to another. Major developed economies were found to be typically net importers of embodied carbon emissions — with UK consumption emissions 34% higher than production emissions, and Germany (29%), Japan (19%) and the USA (13%) also significant net importers of embodied emissions.

Effect of Policy

Governments have taken action to reduce GHG emissions (climate change mitigation). Assessments of policy effectiveness have included work by the Intergovernmental Panel on Climate Change, International Energy Agency, and United Nations Environment Programme. Policies implemented by governments have included national and regional targets to reduce emissions, promoting energy efficiency, and support for renewable energy such as Solar energy as an effective use of renewable energy because solar uses energy from the sun and does not release pollutants into the air.

Countries and regions listed in Annex I of the United Nations Framework Convention on Climate Change (UNFCCC) (i.e., the OECD and former planned economies of the Soviet Union) are required to submit periodic assessments to the UNFCCC of actions they are taking to address climate change. Analysis by the UNFCCC (2011) suggested that policies and measures undertaken by Annex I Parties may have produced emission savings of 1.5 thousand Tg CO_2-eq in the year 2010, with most savings made in the energy sector. The projected emissions saving of 1.5 thousand Tg CO_2-eq is measured against a hypothetical "baseline" of Annex I emissions, i.e., projected Annex I emissions in the absence of policies and measures. The total projected Annex I saving of 1.5 thousand CO_2-eq does not include emissions savings in seven of the Annex I Parties.

Projections

A wide range of projections of future GHG emissions have been produced. Rogner *et al.* (2007) assessed the scientific literature on GHG projections. Rogner *et al.* (2007) concluded that unless energy policies changed substantially, the world would continue to depend on fossil fuels until 2025–2030. Projections suggest that more than 80% of the world's energy will come from fossil fuels. This conclusion was based on "much evidence" and "high agreement" in the literature. Projected annual energy-related CO_2 emissions in 2030 were 40–110% higher than in 2000, with two-thirds of the increase originating in developing countries. Projected annual per capita emissions in developed country regions remained substantially lower (2.8–5.1 tonnes CO_2) than those in developed country regions (9.6–15.1 tonnes CO_2). Projections consistently showed increase in annual world GHG emissions (the "Kyoto" gases, measured in CO_2-equivalent) of 25–90% by 2030, compared to 2000.

Relative CO2 Emission from Various Fuels

One liter of gasoline, when used as a fuel, produces 2.32 kg (about 1300 liters or 1.3 cubic meters) of carbon dioxide, a greenhouse gas. One US gallon produces 19.4 lb (1,291.5 gallons or 172.65 cubic feet)

Mass of carbon dioxide emitted per quantity of energy for various fuels		
Fuel name	**CO_2 emitted (lbs/10^6 Btu)**	**CO_2 emitted (g/MJ)**
Natural gas	117	50.30
Liquefied petroleum gas	139	59.76
Propane	139	59.76
Aviation gasoline	153	65.78
Automobile gasoline	156	67.07
Kerosene	159	68.36
Fuel oil	161	69.22
Tires/tire derived fuel	189	81.26
Wood and wood waste	195	83.83
Coal (bituminous)	205	88.13
Coal (sub-bituminous)	213	91.57
Coal (lignite)	215	92.43
Petroleum coke	225	96.73
Tar-sand Bitumen		
Coal (anthracite)	227	97.59

Life-cycle Greenhouse-gas Emissions of Energy Sources

A literature review of numerous energy sources CO_2 emissions by the IPCC in 2011, found that, the CO_2 emission value that fell within the 50th percentile of all total life cycle emissions studies conducted was as follows.

Lifecycle greenhouse gas emissions by electricity source.		
Technology	**Description**	**50th percentile (g CO_2/ kWh$_e$)**
Hydroelectric	reservoir	4
Ocean Energy	wave and tidal	8
Wind	onshore	12
Nuclear	various generation II reactor types	16
Biomass	various	18
Solar thermal	parabolic trough	22
Geothermal	hot dry rock	45
Solar PV	Polycrystalline silicon	46

Lifecycle greenhouse gas emissions by electricity source.		
Technology	**Description**	**50th percentile (g CO_2/ kWh_e)**
Natural gas	various combined cycle turbines without scrubbing	469
Coal	various generator types without scrubbing	1001

Removal from the Atmosphere ("Sinks")

Natural Processes

Greenhouse gases can be removed from the atmosphere by various processes, as a consequence of:

- a physical change (condensation and precipitation remove water vapor from the atmosphere).

- a chemical reaction within the atmosphere. For example, methane is oxidized by reaction with naturally occurring hydroxyl radical, OH· and degraded to CO_2 and water vapor (CO_2 from the oxidation of methane is not included in the methane Global warming potential). Other chemical reactions include solution and solid phase chemistry occurring in atmospheric aerosols.

- a physical exchange between the atmosphere and the other compartments of the planet. An example is the mixing of atmospheric gases into the oceans.

- a chemical change at the interface between the atmosphere and the other compartments of the planet. This is the case for CO_2, which is reduced by photosynthesis of plants, and which, after dissolving in the oceans, reacts to form carbonic acid and bicarbonate and carbonate ions.

- a photochemical change. Halocarbons are dissociated by UV light releasing Cl· and F· as free radicals in the stratosphere with harmful effects on ozone (halocarbons are generally too stable to disappear by chemical reaction in the atmosphere).

Negative Emissions

A number of technologies remove greenhouse gases emissions from the atmosphere. Most widely analysed are those that remove carbon dioxide from the atmosphere, either to geologic formations such as bio-energy with carbon capture and storage and carbon dioxide air capture, or to the soil as in the case with biochar. The IPCC has pointed out that many long-term climate scenario models require large scale manmade negative emissions to avoid serious climate change.

History of Scientific Research

In the late 19th century scientists experimentally discovered that N2 and O2 do not absorb infrared radiation (called, at that time, "dark radiation"), while water (both as true vapor and condensed in the form of microscopic droplets suspended in clouds) and CO_2 and other poly-atomic gaseous molecules do absorb infrared radiation. In the early 20th century researchers realized that greenhouse gases in the atmosphere made Earth's overall temperature higher than it would be without them. During the late 20th century, a scientific consensus evolved that increasing concentrations of greenhouse gases in the atmosphere cause a substantial rise in global temperatures and changes to other parts of the climate system, with consequences for the environment and for human health.

References

- Jacob, Daniel (1999). Introduction to atmospheric chemistry. Princeton University Press. pp. 25–26. ISBN 0-691-00185-5.

- Wallace, John M. and Peter V. Hobbs. Atmospheric Science; An Introductory Survey.Elsevier. Second Edition, 2006. ISBN 978-0-12-732951-2.

- Bridging the Emissions Gap: A UNEP Synthesis Report (PDF), Nairobi, Kenya: United Nations Environment Programme (UNEP), November 2011, ISBN 978-92-807-3229-0.

- Herzog, T. (November 2006). Yamashita, M. B., ed. Target: intensity — an analysis of greenhouse gas intensity targets (PDF). World Resources Institute. ISBN 1-56973-638-3. Retrieved 2011-04-11.

- World Energy Outlook 2009 (PDF), Paris, France: International Energy Agency (IEA), 2009, pp. 179–180, ISBN 978-92-64-06130-9

- Bridging the Emissions Gap: A UNEP Synthesis Report (PDF), Nairobi, Kenya: United Nations Environment Programme (UNEP), November 2011, ISBN 978-92-807-3229-0.

- St. Fleur, Nicholas (10 November 2015). "Atmospheric Greenhouse Gas Levels Hit Record, Report Says". New York Times. Retrieved 11 November 2015.

- Ritter, Karl (9 November 2015). "UK: In 1st, global temps average could be 1 degree C higher". AP News. Retrieved 11 November 2015.

- Cole, Steve; Gray, Ellen (14 December 2015). "New NASA Satellite Maps Show Human Fingerprint on Global Air Quality". NASA. Retrieved 14 December 2015.

- "search engine optimisation manchester, web design agency london, e commerce manchester". freightbestpractice.org.uk. Retrieved 13 September 2015.

- Mora, C (2013). "The projected timing of climate departure from recent variability". Nature. 502: 183–187. doi:10.1038/nature12540.

- Dumitru-Romulus Târziu; Victor-Dan Păcurar (Jan 2011). "Pădurea, climatul şi energia". Rev. pădur. (in Romanian). 126 (1): 34–39. ISSN 1583-7890. 16720. Retrieved 2012-06-11.

- "Inventory of U.S. Greenhouse Gas Emissions and Sinks: 1990–2010" (PDF). U.S. Environmental Protection Agency. 15 April 2012. p. 1.4. Retrieved 2 June 2012.

- "IEA Publications Bookshop: IEA Publications on 'Energy Policy'". Paris, France: Organization for Economic Co-operation and Development (OECD) / International Energy Agency (IEA). 2012.

- "Greenhouse Gas Emissions from a Typical Passenger Vehicle, US Environment Protection Agency" (PDF). Epa.gov. Retrieved 2011-09-11.

- Bader, N.; Bleichwitz, R. (2009). "Measuring urban greenhouse gas emissions: The challenge of comparability. S.A.P.I.EN.S. 2 (3)". Sapiens.revues.org. Retrieved 2011-09-11.

- Holtz-Eakin, D. (1995). "Stoking the fires? CO_2 emissions and economic growth". Journal of Public Economics. 57 (1): 85–101. doi:10.1016/0047-2727(94)01449-X. Retrieved 2011-04-20.

Types of Greenhouse Gas

The emission of certain toxic gases as a by-product of human activity and their subsequent chemical reaction with particles in the atmosphere has led to those gases being classified as greenhouse gases. These gases radiate energy back to the earth's surface, causing global warming. This chapter lists some of the major greenhouse gases such as carbon dioxide, methane and nitrous oxide.

Carbon Dioxide

Carbon dioxide (chemical formula CO_2) is a colorless and odorless gas vital to life on Earth. This naturally occurring chemical compound is composed of a carbon atom covalently double bonded to two oxygen atoms. Carbon dioxide exists in Earth's atmosphere as a trace gas at a concentration of about 0.04 percent (400 ppm) by volume. Natural sources include volcanoes, hot springs and geysers, and it is freed from carbonate rocks by dissolution in water and acids. Because carbon dioxide is soluble in water, it occurs naturally in groundwater, rivers and lakes, in ice caps and glaciers and also in seawater. It is present in deposits of petroleum and natural gas.

Atmospheric carbon dioxide is the primary source of carbon in life on Earth and its concentration in Earth's pre-industrial atmosphere since late in the Precambrian was regulated by photosynthetic organisms and geological phenomena. As part of the carbon cycle, plants, algae, and cyanobacteria use light energy to photosynthesize carbohydrate from carbon dioxide and water, with oxygen produced as a waste product.

Carbon dioxide (CO_2) is produced by all aerobic organisms when they metabolize carbohydrate and lipids to produce energy by respiration. It is returned to water via the gills of fish and to the air via the lungs of air-breathing land animals, including humans. Carbon dioxide is produced during the processes of decay of organic materials and the fermentation of sugars in bread, beer and winemaking. It is produced by combustion of wood, carbohydrates and fossil fuels such as coal, peat, petroleum and natural gas.

It is a versatile industrial material, used, for example, as an inert gas in welding and fire extinguishers, as a pressurizing gas in air guns and oil recovery, as a chemical feedstock and in liquid form as a solvent in decaffeination of coffee and supercritical drying. It is added to drinking water and carbonated beverages including beer and sparkling wine to add effervescence. The frozen solid form of CO_2, known as "dry ice" is used as a refrigerant and as an abrasive in dry-ice blasting.

Carbon dioxide is an important greenhouse gas. Since the Industrial Revolution, anthropogenic emissions - including the burning of carbon-based fossil fuels and land use changes (primarily deforestation) - have rapidly increased its concentration in the atmosphere, leading to global warming. It is also a major cause of ocean acidification because it dissolves in water to form carbonic acid.

Background

Carbon dioxide was the first gas to be described as a discrete substance. In about 1640, the Flemish chemist Jan Baptist van Helmont observed that when he burned charcoal in a closed vessel, the mass of the resulting ash was much less than that of the original charcoal. His interpretation was that the rest of the charcoal had been transmuted into an invisible substance he termed a "gas" or "wild spirit" (*spiritus sylvestre*).

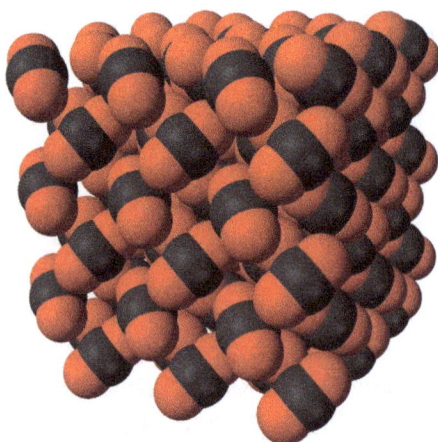

Crystal structure of dry ice

The properties of carbon dioxide were studied more thoroughly in the 1750s by the Scottish physician Joseph Black. He found that limestone (calcium carbonate) could be heated or treated with acids to yield a gas he called "fixed air." He observed that the fixed air was denser than air and supported neither flame nor animal life. Black also found that when bubbled through limewater (a saturated aqueous solution of calcium hydroxide), it would precipitate calcium carbonate. He used this phenomenon to illustrate that carbon dioxide is produced by animal respiration and microbial fermentation. In 1772, English chemist Joseph Priestley published a paper entitled *Impregnating Water with Fixed Air* in which he described a process of dripping sulfuric acid (or *oil of vitriol* as Priestley knew it) on chalk in order to produce carbon dioxide, and forcing the gas to dissolve by agitating a bowl of water in contact with the gas.

Carbon dioxide was first liquefied (at elevated pressures) in 1823 by Humphry Davy and Michael Faraday. The earliest description of solid carbon dioxide was given by Adrien-Jean-Pierre Thilorier, who in 1835 opened a pressurized container of liquid carbon dioxide, only to find that the cooling produced by the rapid evaporation of the liquid yielded a "snow" of solid CO_2.

Chemical and Physical Properties

Structure and Bonding

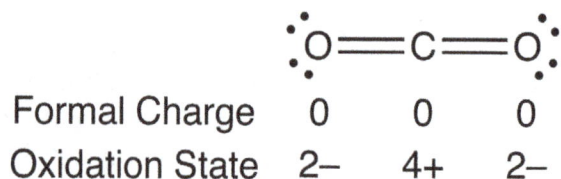

| | Formal Charge | 0 | 0 | 0 |
| | Oxidation State | 2− | 4+ | 2− |

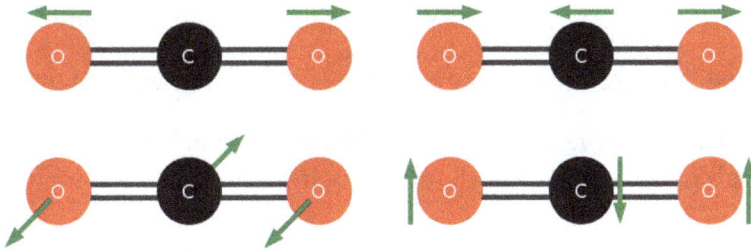

Stretching and bending oscillations of the CO_2 carbon dioxide molecule. Upper left: symmetric stretching. Upper right: antisymmetric stretching. Lower line: degenerate pair of bending modes.

The carbon dioxide molecule is linear and centrosymmetric. The two C=O bonds are equivalent and are short (116.3 pm), consistent with double bonding. Since it is centrosymmetric, the molecule has no electrical dipole. Consequently, only two vibrational bands are observed in the IR spectrum – an antisymmetric stretching mode at 2349 cm⁻¹ and a degenerate pair of bending modes at 667 cm⁻¹. There is also a symmetric stretching mode at 1388 cm⁻¹ which is only observed in the Raman spectrum.

In Aqueous Solution

Carbon dioxide is soluble in water, in which it reversibly forms H_2CO_3 (carbonic acid), which is a weak acid since its ionization in water is incomplete.

$$CO_2 + H_2O \rightleftharpoons H_2CO_3$$

The hydration equilibrium constant of carbonic acid is $K_h = \dfrac{[H_2CO_3]}{[CO_2(aq)]} = 1.70 \times 10^{-3}$ (at 25 °C).

Hence, the majority of the carbon dioxide is not converted into carbonic acid, but remains as CO_2 molecules, not affecting the pH.

The relative concentrations of CO_2, H_2CO_3, and the deprotonated forms HCO_3^- (bicarbonate) and CO2−3(carbonate) depend on the pH. As shown in a Bjerrum plot, in neutral or slightly alkaline water (pH > 6.5), the bicarbonate form predominates (>50%) becoming the most prevalent (>95%) at the pH of seawater. In very alkaline water (pH > 10.4), the predominant (>50%) form is carbonate. The oceans, being mildly alkaline with typical pH = 8.2–8.5, contain about 120 mg of bicarbonate per liter.

Being diprotic, carbonic acid has two acid dissociation constants, the first one for the dissociation into the bicarbonate (also called hydrogen carbonate) ion (HCO_3^-):

$$H_2CO_3 \rightleftharpoons HCO_3^- + H^+$$

$K_{a1} = 2.5 \times 10^{-4}$ mol/L; $pK_{a1} = 3.6$ at 25 °C.

This is the *true* first acid dissociation constant, defined as $K_{a1} = \dfrac{[HCO_3^-][H^+]}{[H_2CO_3]}$, where the denominator includes only covalently bound H_2CO_3 and does not include hydrated $CO_2(aq)$. The much smaller and often-quoted value near 4.16×10^{-7} is an *apparent* value calculated on the (incorrect) assumption that all dissolved CO_2 is present as carbonic acid, so that

$K_{a1}(\text{apparent}) = \dfrac{[HCO_3^-][H^+]}{[H_2CO_3]+[CO_2(aq)]}$. Since most of the dissolved CO_2 remains as CO_2 molecules, $K_{a1}(\text{apparent})$ has a much larger denominator and a much smaller value than the true K_{a1}.

The bicarbonate ion is an amphoteric species that can act as an acid or as a base, depending on pH of the solution. At high pH, it dissociates significantly into the carbonate ion (CO_3^{2-}):

$HCO_3^- \rightleftharpoons CO_3^{2-} + H^+$

$K_{a2} = 4.69 \times 10^{-11}$ mol/L; p$K_{a2} = 10.329$

In organisms carbonic acid production is catalysed by the enzyme, carbonic anhydrase.

Chemical Reactions of CO_2

CO_2 is a weak electrophile. Its reaction with basic water illustrates this property, in which case hydroxide is the nucleophile. Other nucleophiles react as well. For example, carbanions as provided by Grignard reagents and organolithium compounds react with CO_2 to give carboxylates:

$MR + CO_2 \rightarrow RCO_2M$

where M = Li or MgBr and R = alkyl or aryl.

In metal carbon dioxide complexes, CO_2 serves as a ligand, which can facilitate the conversion of CO_2 to other chemicals.

The reduction of CO_2 to CO is ordinarily a difficult and slow reaction:

$CO_2 + 2\,e^- + 2H^+ \rightarrow CO + H_2O$

The redox potential for this reaction near pH 7 is about −0.53 V *versus* the standard hydrogen electrode. The nickel-containing enzyme carbon monoxide dehydrogenase catalyses this process.

Physical Properties

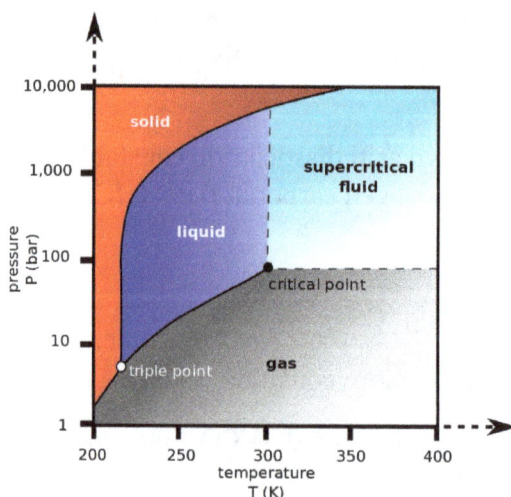

Carbon dioxide pressure-temperature phase diagram showing the triple point and critical point of carbon dioxide

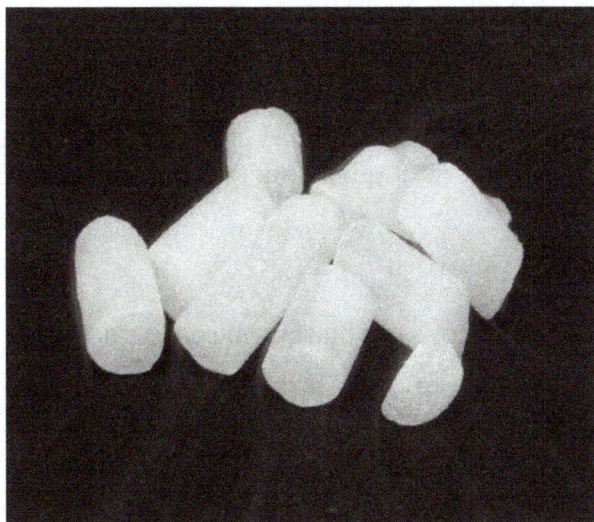

Sample of solid carbon dioxide or "dry ice" pellets

Carbon dioxide is colorless. At low concentrations, the gas is odorless. At higher concentrations it has a sharp, acidic odor. At standard temperature and pressure, the density of carbon dioxide is around 1.98 kg/m³, about 1.67 times that of air.

Carbon dioxide has no liquid state at pressures below 5.1 standard atmospheres (520 kPa). At 1 atmosphere (near mean sea level pressure), the gas deposits directly to a solid at temperatures below −78.5 °C (−109.3 °F; 194.7 K) and the solid sublimes directly to a gas above −78.5 °C. In its solid state, carbon dioxide is commonly called dry ice.

Liquid carbon dioxide forms only at pressures above 5.1 atm; the triple point of carbon dioxide is about 518 kPa at −56.6 °C . The critical point is 7.38 MPa at 31.1 °C. Another form of solid carbon dioxide observed at high pressure is an amorphous glass-like solid. This form of glass, called *carbonia*, is produced by supercooling heated CO_2 at extreme pressure (40–48 GPa or about 400,000 atmospheres) in a diamond anvil. This discovery confirmed the theory that carbon dioxide could exist in a glass state similar to other members of its elemental family, like silicon (silica glass) and germanium dioxide. Unlike silica and germania glasses, however, carbonia glass is not stable at normal pressures and reverts to gas when pressure is released.

At temperatures and pressures above the critical point, carbon dioxide behaves as a supercritical fluid known as supercritical carbon dioxide.

Isolation and Production

Carbon dioxide can be obtained by distillation from air, but the method is inefficient. Industrially, carbon dioxide is predominantly an unrecovered waste product, produced by several methods which may be practiced at various scales.

The combustion of all carbon-based fuels, such as methane (natural gas), petroleum distillates (gasoline, diesel, kerosene, propane), coal, wood and generic organic matter produces carbon dioxide and, except in the case of pure carbon, water. As an example, the chemical reaction between methane and oxygen is given below.

$$CH_4 + 2 O_2 \rightarrow CO_2 + 2 H_2O$$

It is produced by thermal decomposition of limestone, $CaCO_3$ by heating (calcining) at about 850 °C (1,560 °F), in the manufacture of quicklime (calcium oxide, CaO), a compound that has many industrial uses:

$$CaCO_3 \rightarrow CaO + CO_2$$

Iron is reduced from its oxides with coke in a blast furnace, producing pig iron and carbon dioxide:

Carbon dioxide is a byproduct of the industrial production of hydrogen by steam reforming and ammonia synthesis. These processes begin with the reaction of water and natural gas (mainly methane).

Acids liberate CO_2 from most metal carbonates. Consequently, it may be obtained directly from natural carbon dioxide springs, where it is produced by the action of acidified water on limestone or dolomite. The reaction between hydrochloric acid and calcium carbonate (limestone or chalk) is shown below:

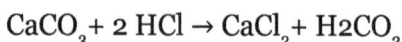

$$CaCO_3 + 2 HCl \rightarrow CaCl_2 + H2CO_3$$

The carbonic acid (H2CO3) then decomposes to water and CO_2:

$$H_2CO_3 \rightarrow CO_2 + H_2O$$

Such reactions are accompanied by foaming or bubbling, or both, as the gas is released. They have widespread uses in industry because they can be used to neutralize waste acid streams.

Carbon dioxide is a by-product of the fermentation of sugar in the brewing of beer, whisky and other alcoholic beverages and in the production of bioethanol. Yeast metabolizes sugar to produce CO_2 and ethanol, also known as alcohol, as follows:

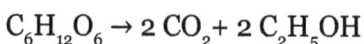

$$C_6H_{12}O_6 \rightarrow 2 CO_2 + 2 C_2H_5OH$$

All aerobic organisms produce CO_2 when they oxidize carbohydrates, fatty acids, and proteins. The large number of reactions involved are exceedingly complex and not described easily. Refer to (cellular respiration, anaerobic respiration and photosynthesis). The equation for the respiration of glucose and other monosaccharides is:

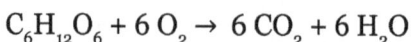

$$C_6H_{12}O_6 + 6 O_2 \rightarrow 6 CO_2 + 6 H_2O$$

Photoautotrophs (i.e. plants and cyanobacteria) use the energy contained in sunlight to photosynthesize simple sugars from CO_2 absorbed from the air and water:

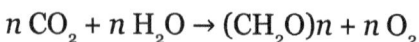

$$n CO_2 + n H_2O \rightarrow (CH_2O)n + n O_2$$

Carbon dioxide comprises about 40-45% of the gas that emanates from decomposition in landfills (termed "landfill gas"). Most of the remaining 50-55% is methane.

Uses

Carbon dioxide is used by the food industry, the oil industry, and the chemical industry. The compound has varied commercial uses but one of its greatest use as a chemical is in the production of carbonated beverages; it provides the sparkle in carbonated beverages such as soda water.

Carbon dioxide bubbles in a soft drink.

Precursor to Chemicals

In the chemical industry, carbon dioxide is mainly consumed as an ingredient in the production of urea, with a smaller fraction being used to produce methanol and a range of other products. Metal carbonates and bicarbonates, as well as some carboxylic acids derivatives (e.g., sodium salicylate) are prepared using CO_2 by the Kolbe-Schmitt reaction.

In addition to conventional processes using CO_2 for chemical production, electrochemical methods are also being explored at a research level. In particular, the use of renewable energy for production of fuels from CO_2 (such as methanol) is attractive as this could result in fuels that could be easily transported and used within conventional combustion technologies but have no net CO_2 emissions.

Foods

Carbon dioxide is a food additive used as a propellant and acidity regulator in the food industry. It is approved for usage in the EU (listed as E number E290), US and Australia and New Zealand (listed by its INS number 290).

A candy called Pop Rocks is pressurized with carbon dioxide gas at about 4×10^6 Pa (40 bar, 580 psi). When placed in the mouth, it dissolves (just like other hard candy) and releases the gas bubbles with an audible pop.

Leavening agents cause dough to rise by producing carbon dioxide. Baker's yeast produces carbon dioxide by fermentation of sugars within the dough, while chemical leaveners such as baking powder and baking soda release carbon dioxide when heated or if exposed to acids.

Beverages

Carbon dioxide is used to produce carbonated soft drinks and soda water. Traditionally, the carbonation of beer and sparkling wine came about through natural fermentation, but many manufacturers carbonate these drinks with carbon dioxide recovered from the fermentation process. In the case of bottled and kegged beer, the most common method used is carbonation with recycled carbon dioxide. With the exception of British Real Ale, draught beer is usually transferred from kegs in a cold room or cellar to dispensing taps on the bar using pressurized carbon dioxide, sometimes mixed with nitrogen.

Wine Making

Carbon dioxide in the form of dry ice is often used in the wine making process to cool down bunches of grapes quickly after picking to help prevent spontaneous fermentation by wild yeast. The main advantage of using dry ice over regular water ice is that it cools the grapes without adding any additional water that may decrease the sugar concentration in the grape must, and therefore also decrease the alcohol concentration in the finished wine.

Dry ice is also used during the cold soak phase of the wine making process to keep grapes cool. The carbon dioxide gas that results from the sublimation of the dry ice tends to settle to the bottom of tanks because it is denser than air. The settled carbon dioxide gas creates a hypoxic environment which helps to prevent bacteria from growing on the grapes until it is time to start the fermentation with the desired strain of yeast.

Carbon dioxide is also used to create a hypoxic environment for carbonic maceration, the process used to produce Beaujolais wine.

Carbon dioxide is sometimes used to top up wine bottles or other storage vessels such as barrels to prevent oxidation, though it has the problem that it can dissolve into the wine, making a previously still wine slightly fizzy. For this reason, other gases such as nitrogen or argon are preferred for this process by professional wine makers.

Inert Gas

It is one of the most commonly used compressed gases for pneumatic (pressurized gas) systems in portable pressure tools. Carbon dioxide is also used as an atmosphere for welding, although in the welding arc, it reacts to oxidize most metals. Use in the automotive industry is common despite significant evidence that welds made in carbon dioxide are more brittle than those made in more inert atmospheres, and that such weld joints deteriorate over time because of the formation of carbonic acid. It is used as a welding gas primarily because it is much less expensive than more inert gases such as argon or helium. When used for MIG welding, CO_2 use is sometimes referred to as MAG welding, for Metal Active Gas, as CO_2 can react at these high temperatures. It tends to produce a hotter puddle than truly inert atmospheres, improving the flow characteristics. Although, this may be due to atmospheric reactions occurring at the puddle site. This is usually the opposite of the desired effect when welding, as it tends to embrittle the site, but may not be a problem for general mild steel welding, where ultimate ductility is not a major concern.

It is used in many consumer products that require pressurized gas because it is inexpensive and nonflammable, and because it undergoes a phase transition from gas to liquid at room temperature at an attainable pressure of approximately 60 bar (870 psi, 59 atm), allowing far more carbon dioxide to fit in a given container than otherwise would. Life jackets often contain canisters of pressured carbon dioxide for quick inflation. Aluminium capsules of CO_2 are also sold as supplies of compressed gas for airguns, paintball markers, inflating bicycle tires, and for making carbonated water. Rapid vaporization of liquid carbon dioxide is used for blasting in coal mines. High concentrations of carbon dioxide can also be used to kill pests. Liquid carbon dioxide is used in supercritical drying of some food products and technological materials, in the preparation of specimens for scanning electron microscopy and in the decaffeination of coffee beans.

Fire Extinguisher

Carbon dioxide can be used to extinguishes flames by flooding the environment around the flame with the gas. It does not itself react to extinguish the flame, but starves the flame of oxygen by displacing it. Some fire extinguishers, especially those designed for electrical fires, contain liquid carbon dioxide under pressure. Carbon dioxide extinguishers work well on small flammable liquid and electrical fires, but not on ordinary combustible fires, because although it excludes oxygen, it does not cool the burning substances significantly and when the carbon dioxide disperses they are free to catch fire upon exposure to atmospheric oxygen. Their desirability in electrical fire stems from the fact that, unlike water or other chemical based methods, Carbon dioxide will not cause short circuits, leading to even more damage to equipment. Because it is a gas, it is also easy to dispense large amounts of the gas automatically in IT infrastructure rooms, where the fire itself might be hard to reach with more immediate methods because it is behind rack doors and inside of cases. Carbon dioxide has also been widely used as an extinguishing agent in fixed fire protection systems for local application of specific hazards and total flooding of a protected space. International Maritime Organization standards also recognize carbon dioxide systems for fire protection of ship holds and engine rooms. Carbon dioxide based fire protection systems have been linked to several deaths, because it can cause suffocation in sufficiently high concentrations. A review of CO_2 systems identified 51 incidents between 1975 and the date of the report, causing 72 deaths and 145 injuries.

Supercritical CO_2 as Solvent

Liquid carbon dioxide is a good solvent for many lipophilic organic compounds and is used to remove caffeine from coffee. Carbon dioxide has attracted attention in the pharmaceutical and other chemical processing industries as a less toxic alternative to more traditional solvents such as organochlorides. It is used by some dry cleaners for this reason. It is used in the preparation of some aerogels because of the properties of supercritical carbon dioxide.

Agricultural and Biological Applications

Plants require carbon dioxide to conduct photosynthesis. The atmospheres of greenhouses may (if of large size, must) be enriched with additional CO_2 to sustain and increase the rate of plant growth. At very high concentrations (100 times atmospheric concentration, or greater), carbon dioxide can be toxic to animal life, so raising the concentration to 10,000 ppm (1%) or higher for several hours will eliminate pests such as whiteflies and spider mites in a greenhouse.

In medicine, up to 5% carbon dioxide (130 times atmospheric concentration) is added to oxygen for stimulation of breathing after apnea and to stabilize the O_2/CO_2 balance in blood.

It has been proposed that carbon dioxide from power generation be bubbled into ponds to stimulate growth of algae that could then be converted into biodiesel fuel.

Oil Recovery

Carbon dioxide is used in enhanced oil recovery where it is injected into or adjacent to producing oil wells, usually under supercritical conditions, when it becomes miscible with the oil. This approach can increase original oil recovery by reducing residual oil saturation by between 7 per cent to 23 per cent additional to primary extraction. It acts as both a pressurizing agent and, when dissolved into the underground crude oil, significantly reduces its viscosity, and changing surface chemistry enabling the oil to flow more rapidly through the reservoir to the removal well. In mature oil fields, extensive pipe networks are used to carry the carbon dioxide to the injection points.

Bio Transformation Into Fuel

Researchers have genetically modified a strain of the cyanobacterium *Synechococcus elongatus* to produce the fuels isobutyraldehyde and isobutanol from CO_2 using photosynthesis.

Refrigerant

Liquid and solid carbon dioxide are important refrigerants, especially in the food industry, where they are employed during the transportation and storage of ice cream and other frozen foods. Solid carbon dioxide is called "dry ice" and is used for small shipments where refrigeration equipment is not practical. Solid carbon dioxide is always below −78.5 °C at regular atmospheric pressure, regardless of the air temperature.

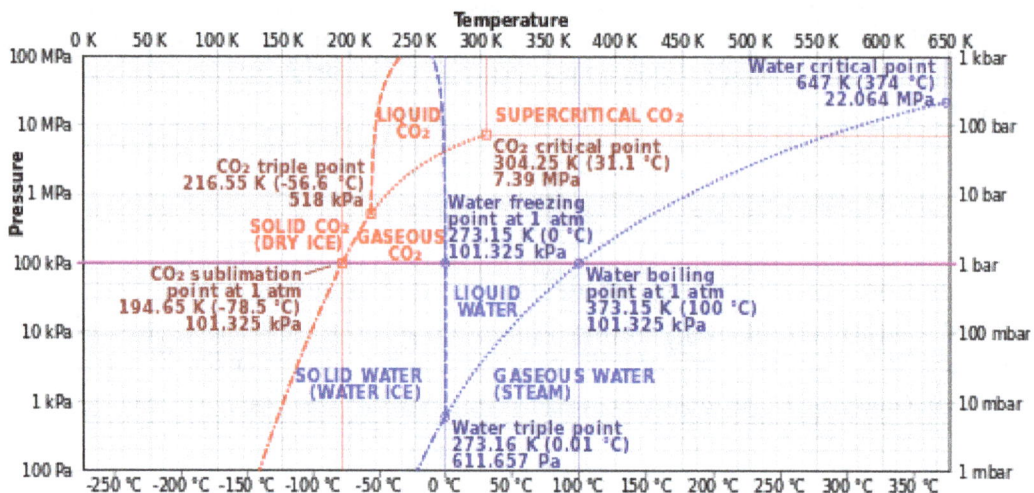

Comparison of phase diagrams of carbon dioxide (red) and water (blue) as a log-lin chart with phase transitions points at 1 atmosphere

Liquid carbon dioxide (industry nomenclature R744 or R-744) was used as a refrigerant prior to the discovery of R-12 and may enjoy a renaissance due to the fact that R134a contributes to climate change. Its physical properties are highly favorable for cooling, refrigeration, and heating purposes, having a high volumetric cooling capacity. Due to the need to operate at pressures of up to 130 bar (1880 psi), CO_2 systems require highly resistant components that have already been developed for mass production in many sectors. In automobile air conditioning, in more than 90% of all driving conditions for latitudes higher than 50°, R744 operates more efficiently than systems using R134a. Its environmental advantages (GWP of 1, non-ozone depleting, non-toxic, non-flammable) could make it the future working fluid to replace current HFCs in cars, supermarkets, and heat pump water heaters, among others. Coca-Cola has fielded CO_2-based beverage coolers and the U.S. Army is interested in CO_2 refrigeration and heating technology.

The global automobile industry is expected to decide on the next-generation refrigerant in car air conditioning. CO_2 is one discussed option.

Coal Bed Methane Recovery

In enhanced coal bed methane recovery, carbon dioxide would be pumped into the coal seam to displace methane, as opposed to current methods which primarily rely on the removal of water (to reduce pressure) to make the coal seam release its trapped methane.

Niche Uses

Carbon dioxide is the lasing medium in a carbon dioxide laser, which is one of the earliest type of lasers.

A carbon dioxide laser.

Carbon dioxide can be used as a means of controlling the pH of swimming pools, by continuously adding gas to the water, thus keeping the pH from rising. Among the advantages of this is the avoidance of handling (more hazardous) acids. Similarly, it is also used in the maintaining reef aquaria, where it is commonly used in calcium reactors to temporarily lower the pH of water being passed over calcium carbonate in order to allow the calcium carbonate to dissolve into the water more freely where it is used by some corals to build their skeleton.

Used as the primary coolant in the British advanced gas-cooled reactor for nuclear power generation.

Carbon dioxide induction is commonly used for the euthanasia of laboratory research animals. Methods to administer CO_2 include placing animals directly into a closed, prefilled chamber containing CO_2, or exposure to a gradually increasing concentration of CO_2. In 2013, the American Veterinary Medical Association issued new guidelines for carbon dioxide induction, stating that a displacement rate of 10% to 30% of the gas chamber volume per minute is optimal for the humane euthanization of small rodents.

Carbon dioxide is also used in several related cleaning and surface preparation techniques.

In Earth's Atmosphere

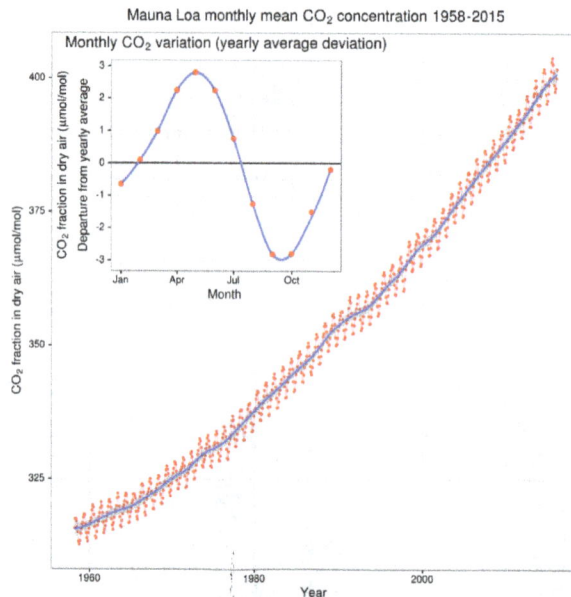

The Keeling Curve of atmospheric CO_2 concentrations measured at Mauna Loa Observatory

Carbon dioxide in Earth's atmosphere is a trace gas, currently (early 2016) having an average concentration of 402 parts per million by volume (or 611 parts per million by mass). Atmospheric concentrations of carbon dioxide fluctuate slightly with the seasons, falling during the Northern Hemisphere spring and summer as plants consume the gas and rising during northern autumn and winter as plants go dormant or die and decay. Concentrations also vary on a regional basis, most strongly near the ground with much smaller variations aloft. In urban areas concentrations are generally higher and indoors they can reach 10 times background levels.

Combustion of fossil fuels and deforestation have caused the atmospheric concentration of carbon dioxide to increase by about 43% since the beginning of the age of industrialization. Most carbon dioxide from human activities is released from burning coal and other fossil fuels. Other human activities, including deforestation, biomass burning, and cement production also produce carbon dioxide. Volcanoes emit between 0.2 and 0.3 billion tons of carbon dioxide per year, while human activities emit about 29 billion tons.

Carbon dioxide is a greenhouse gas, absorbing and emitting infrared radiation at its two infrared-active vibrational frequencies. This process causes carbon dioxide to warm the surface and lower atmosphere, while cooling the upper atmosphere. The increase in atmospheric concentration of CO_2, and thus in the CO_2-induced greenhouse effect, is the reason for the rise in average global temperature since the mid-20th century. Although carbon dioxide is the greenhouse gas primarily responsible for the rise, methane, nitrous oxide, ozone, and various other long-lived greenhouse gases also contribute. Carbon dioxide is of greatest concern because it exerts a larger overall warming influence than all of those other gases combined, and because it has a long atmospheric lifetime.

Not only do increasing carbon dioxide concentrations lead to increases in global surface temperature, but increasing global temperatures also cause increasing concentrations of carbon dioxide. This produces a positive feedback for changes induced by other processes such as orbital cycles. Five hundred million years ago the carbon dioxide concentration was 20 times greater than today, decreasing to 4–5 times during the Jurassic period and then slowly declining with a particularly swift reduction occurring 49 million years ago.

Local concentrations of carbon dioxide can reach high values near strong sources, especially those that are isolated by surrounding terrain. At the Bossoleto hot spring near Rapolano Terme in Tuscany, Italy, situated in a bowl-shaped depression about 100 m (330 ft) in diameter, concentrations of CO_2 rise to above 75% overnight, sufficient to kill insects and small animals. After sunrise the gas is dispersed by convection during the day. High concentrations of CO_2 produced by disturbance of deep lake water saturated with CO_2 are thought to have caused 37 fatalities at Lake Monoun, Cameroon in 1984 and 1700 casualties at Lake Nyos, Cameroon in 1986.

On November 12, 2015, NASA scientists reported that human-made carbon dioxide (CO_2) continues to increase above levels not seen in hundreds of thousands of years: currently, about half of the carbon dioxide released from the burning of fossil fuels remains in the atmosphere and is not absorbed by vegetation and the oceans.

In the Oceans

Carbon dioxide dissolves in the ocean to form carbonic acid (H_2CO_3), bicarbonate (HCO_3^-) and carbonate (CO_3^{2-}). There is about fifty times as much carbon dissolved in the oceans as exists in the atmosphere. The oceans act as an enormous carbon sink, and have taken up about a third of CO_2 emitted by human activity.

As the concentration of carbon dioxide increases in the atmosphere, the increased uptake of carbon dioxide into the oceans is causing a measurable decrease in the pH of the oceans, which is referred to as ocean acidification. This reduction in pH affects biological systems in the oceans, primarily oceanic calcifying organisms. These effects span the food chain from autotrophs to heterotrophs and include organisms such as coccolithophores, corals, foraminifera, echinoderms, crustaceans and mollusks. Under normal conditions, calcium carbonate is stable in surface waters since the carbonate ion is at supersaturating concentrations. However, as ocean pH falls, so does the concentration of this ion, and when carbonate becomes undersaturated, structures made of calcium carbonate are vulnerable to dissolution. Corals, coccolithophore algae, coralline algae, foraminifera, shellfish and pteropods experience reduced calcification or enhanced dissolution when exposed to elevated CO_2.

Gas solubility decreases as the temperature of water increases (except when both pressure exceeds 300 bar and temperature exceeds 393 K, only found near deep geothermal vents) and therefore the rate of uptake from the atmosphere decreases as ocean temperatures rise.

Most of the CO_2 taken up by the ocean, which is about 30% of the total released into the atmosphere, forms carbonic acid in equilibrium with bicarbonate. Some of these chemical species are consumed by photosynthetic organisms that remove carbon from the cycle. Increased CO_2 in the atmosphere has led to decreasing alkalinity of seawater, and there is concern that this may adversely affect organisms living in the water. In particular, with decreasing alkalinity, the availability of carbonates for forming shells decreases, although there's evidence of increased shell production by certain species under increased CO_2 content.

NOAA states in their May 2008 "State of the science fact sheet for ocean acidification" that: "The oceans have absorbed about 50% of the carbon dioxide (CO_2) released from the burning of fossil fuels, resulting in chemical reactions that lower ocean pH. This has caused an increase in hydrogen ion (acidity) of about 30% since the start of the industrial age through a process known as "ocean acidification." A growing number of studies have demonstrated adverse impacts on marine organisms, including:

- The rate at which reef-building corals produce their skeletons decreases, while production of numerous varieties of jellyfish increases.

- The ability of marine algae and free-swimming zooplankton to maintain protective shells is reduced.

- The survival of larval marine species, including commercial fish and shellfish, is reduced."

Also, the Intergovernmental Panel on Climate Change (IPCC) writes in their Climate Change 2007: Synthesis Report: "The uptake of anthropogenic carbon since 1750 has led to the ocean becoming more acidic with an average decrease in pH of 0.1 units. Increasing atmospheric CO_2 concentrations lead to further acidification ... While the effects of observed ocean acidification on the marine biosphere are as yet undocum ented, the progressive acidification of oceans is expected to have negative impacts on marine shell-forming organisms (e.g. corals) and their dependent species."

Some marine calcifying organisms (including coral reefs) have been singled out by major research agencies, including NOAA, OSPAR commission, NANOOS and the IPCC, because their most current research shows that ocean acidification should be expected to impact them negatively.

Carbon dioxide is also introduced into the oceans through hydrothermal vents. The *Champagne* hydrothermal vent, found at the Northwest Eifuku volcano at Marianas Trench Marine National-al Monument, produces almost pure liquid carbon dioxide, one of only two known sites in the world.

Biological Role

Carbon dioxide is an end product of cellular respiration in organisms that obtain energy by breaking down sugars, fats and amino acids with oxygen as part of their metabolism. This includes all plants, algae and animals and aerobic fungi and bacteria. In vertebrates, the carbon dioxide travels in the blood from the body's tissues to the skin (e.g., amphibians) or the gills (e.g., fish), from where it dissolves in the water, or to the lungs from where it is exhaled. During active photosynthesis, plants can absorb more carbon dioxide from the atmosphere than they release in respiration.

Photosynthesis and Carbon Fixation

Carbon fixation is a biochemical process by which atmospheric carbon dioxide is incorporated by plants, algae and (cyanobacteria) into energy-rich organic molecules such as glucose, thus creating their own food by photosynthesis. Photosynthesis uses carbon dioxide and water to produce sugars from which other organic compounds can be constructed, and oxygen is produced as a by-product.

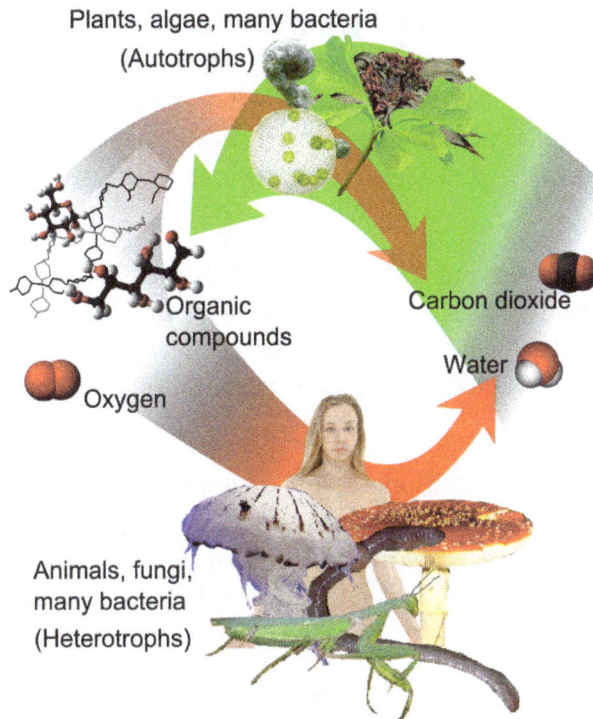

Overview of photosynthesis and respiration. Carbon dioxide (at right), together with water, form oxygen and organic compounds (at left) by photosynthesis, which can be respired to water and (CO_2).

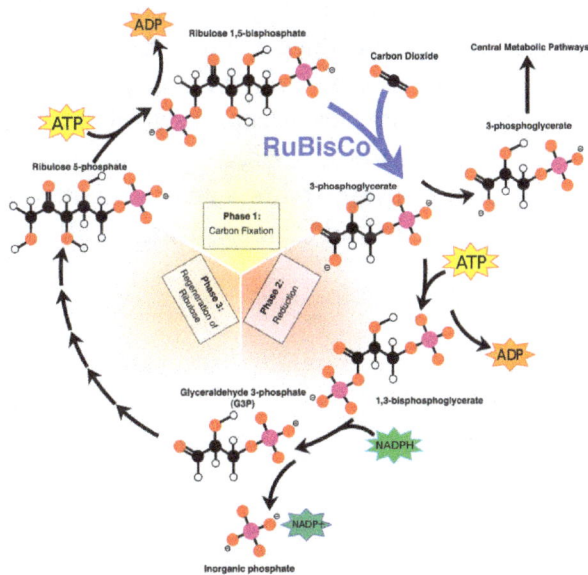

Figure Overview of the Calvin cycle and carbon fixation

Ribulose-1,5-bisphosphate carboxylase oxygenase, commonly abbreviated to RuBisCO, is the enzyme involved in the first major step of carbon fixation, the production of two molecules of 3-phos-phoglycerate from CO_2 and ribulose bisphosphate, as shown in the diagram.

RuBisCO is thought to be the single most abundant protein on Earth.

Phototrophs use the products of their photosynthesis as internal food sources and as raw material for the biosynthesis of more complex organic molecules, such as polysaccharides, nucleic acids and proteins. These are used for their own growth, and also as the basis of the food chains and webs that feed other organisms, including animals such as ourselves. Some important phototrophs, the coccolithophores synthesise hard calcium carbonate scales. A globally significant species of coccolithophore is *Emiliania huxleyi* whose calcite scales have formed the basis of many sedimentary rocks such as limestone, where what was previously atmospheric carbon can remain fixed for geological timescales.

Plants can grow as much as 50 percent faster in concentrations of 1,000 ppm CO_2 when compared with ambient conditions, though this assumes no change in climate and no limitation on other nutrients. Elevated CO_2 levels cause increased growth reflected in the harvestable yield of crops, with wheat, rice and soybean all showing increases in yield of 12–14% under elevated CO_2 in FACE experiments.

Increased atmospheric CO_2 concentrations result in fewer stomata developing on plants which leads to reduced water usage and increased water-use efficiency. Studies using FACE have shown that CO_2 enrichment leads to decreased concentrations of micronutrients in crop plants. This may have knock-on effects on other parts of ecosystems as herbivores will need to eat more food to gain the same amount of protein.

The concentration of secondary metabolites such as phenylpropanoids and flavonoids can also be altered in plants exposed to high concentrations of CO_2.

Plants also emit CO_2 during respiration, and so the majority of plants and algae, which use C3 photosynthesis, are only net absorbers during the day. Though a growing forest will absorb many tons of CO_2 each year, a mature forest will produce as much CO_2 from respiration and decomposition of dead specimens (e.g., fallen branches) as is used in photosynthesis in growing plants. Contrary to the long-standing view that they are carbon neutral, mature forests can continue to accumulate carbon and remain valuable carbon sinks, helping to maintain the carbon balance of Earth's atmosphere. Additionally, and crucially to life on earth, photosynthesis by phytoplankton consumes dissolved CO_2 in the upper ocean and thereby promotes the absorption of CO_2 from the atmosphere.

Toxicity

Carbon dioxide content in fresh air (averaged between sea-level and 10 kPa level, i.e., about 30 km (19 mi) altitude) varies between 0.036% (360 ppm) and 0.041% (410 ppm), depending on the location.

Main symptoms of
Carbon dioxide toxicity

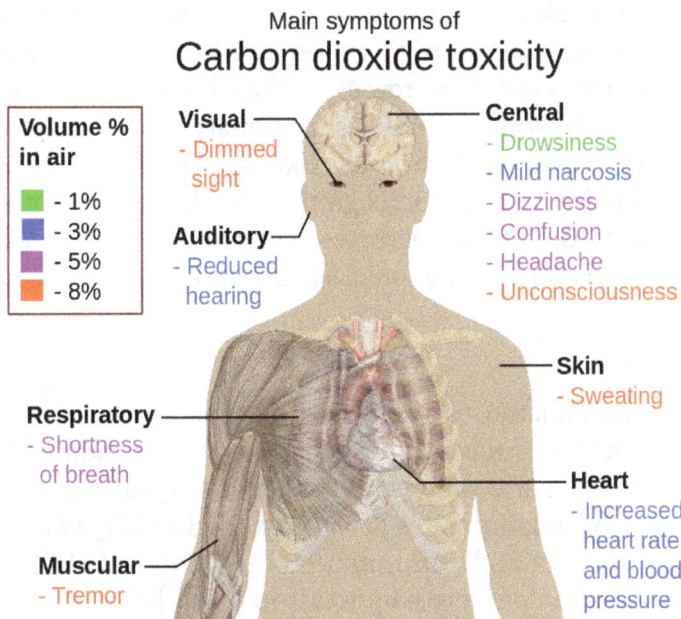

Main symptoms of carbon dioxide toxicity, by increasing volume percent in air.

CO_2 is an asphyxiant gas and not classified as toxic or harmful in accordance with Globally Harmonized System of Classification and Labelling of Chemicals standards of United Nations Economic Commission for Europe by using the OECD Guidelines for the Testing of Chemicals. In concentrations up to 1% (10,000 ppm), it will make some people feel drowsy and give the lungs a stuffy feeling. Concentrations of 7% to 10% (70,000 to 100,000 ppm) may cause suffocation, even in the presence of sufficient oxygen, manifesting as dizziness, headache, visual and hearing dysfunction, and unconsciousness within a few minutes to an hour. The physiological effects of acute carbon dioxide exposure are grouped together under the term hypercapnia, a subset of asphyxiation.

Because it is heavier than air, in locations where the gas seeps from the ground (due to sub-surface volcanic or geothermal activity) in relatively high concentrations, without the dispersing effects of wind, it can collect in sheltered/pocketed locations below average ground level, causing animals located therein to be suffocated. Carrion feeders attracted to the carcasses are then also killed. Children have been killed in the same way near the city of Goma by CO_2 emissions from the nearby volcano Mt. Nyiragongo. The Swahili term for this phenomenon is 'mazuku'.

Adaptation to increased concentrations of CO_2 occurs in humans, including modified breathing and kidney bicarbonate production, in order to balance the effects of blood acidification (acidosis). Several studies suggested that 2.0 percent inspired concentrations could be used for closed air spaces (e.g. a submarine) since the adaptation is physiological and reversible, as decrement in performance or in normal physical activity does not happen at this level of exposure for five days. Yet, other studies show a decrease in cognitive function even at much lower levels. Also, with ongoing respiratory acidosis, adaptation or compensatory mechanisms will be unable to reverse such condition.

Below 1%

There are few studies of the health effects of long-term continuous CO_2 exposure on humans and animals at levels below 1% and there is potentially a significant risk to humans in the near future with rising atmospheric CO_2 levels associated with climate change. Occupational CO_2 exposure limits have been set in the United States at 0.5% (5000 ppm) for an eight-hour period. At this CO_2 concentration, International Space Station crew experienced headaches, lethargy, mental slowness, emotional irritation, and sleep disruption. Studies in animals at 0.5% CO_2 have demonstrated kidney calcification and bone loss after eight weeks of exposure. A study of humans exposed in 2.5 hour sessions demonstrated significant effects on cognitive abilities at concentrations as low as 0.1% (1000ppm) CO_2 likely due to CO_2 induced increases in cerebral blood flow. Another study observed a decline in basic activity level and information usage at 1000 ppm, when compared to 500 ppm.

Ventilation

Poor ventilation is one of the main causes of excessive CO_2 concentrations in closed spaces. Carbon dioxide differential above outdoor concentrations at steady state conditions (when the occupancy and ventilation system operation are sufficiently long that CO_2 concentration has stabilized) are sometimes used to estimate ventilation rates per person. Higher CO_2 concentrations are associated with occupant health, comfort and performance degradation. ASHRAE Standard 62.1–2007 ventilation rates may result in indoor levels up to 2,100 ppm above ambient outdoor conditions. Thus if the outdoor concentration is 400 ppm, indoor concentrations may reach 2,500 ppm with ventilation rates that meet this industry consensus standard. Concentrations in poorly ventilated spaces can be found even higher than this (range of 3,000 or 4,000).

Miners, who are particularly vulnerable to gas exposure due to an insufficient ventilation, referred to mixtures of carbon dioxide and nitrogen as "blackdamp," "choke damp" or "stythe." Before more effective technologies were developed, miners would frequently monitor for dangerous levels of blackdamp and other gases in mine shafts by bringing a caged canary with them as they worked. The canary is more sensitive to asphyxiant gases than humans, and as it became unconscious would stop singing and fall off its perch. The Davy lamp could also detect high levels of blackdamp (which sinks, and collects near the floor) by burning less brightly, while methane, another suffocating gas and explosion risk, would make the lamp burn more brightly.

Human Physiology

Content

The body produces approximately 2.3 pounds (1.0 kg) of carbon dioxide per day per person, containing 0.63 pounds (290 g) of carbon. In humans, this carbon dioxide is carried through the venous system and is breathed out through the lungs, resulting in lower concentrations in the arteries. The carbon dioxide content of the blood is often given as the partial pressure, which is the pressure which carbon dioxide would have had if it alone occupied the volume.

In humans, the carbon dioxide contents are as follows:

Reference ranges or averages for partial pressures of carbon dioxide (abbreviated PCO_2)			
Unit	Venous blood gas	Alveolar pulmonary gas pressures	Arterial blood carbon dioxide
kPa	5.5-6.8	4.8	4.7-6.0
mmHg	41–51	36	35-45

Transport in the Blood

CO_2 is carried in blood in three different ways. (The exact percentages vary depending whether it is arterial or venous blood).

Most of it (about 70% to 80%) is converted to bicarbonate ions HCO−3 by the enzyme carbonic anhydrase in the red blood cells, by the reaction $CO_2 + H_2O \rightarrow H_2CO_3 \rightarrow H^+ + HCO−3$.

- 5% – 10% is dissolved in the plasma
- 5% – 10% is bound to hemoglobin as carbamino compounds

Hemoglobin, the main oxygen-carrying molecule in red blood cells, carries both oxygen and carbon dioxide. However, the CO_2 bound to hemoglobin does not bind to the same site as oxygen. Instead, it combines with the N-terminal groups on the four globin chains. However, because of allosteric effects on the hemoglobin molecule, the binding of CO_2 decreases the amount of oxygen that is bound for a given partial pressure of oxygen. The decreased binding to carbon dioxide in the blood due to increased oxygen levels is known as the Haldane Effect, and is important in the transport of carbon dioxide from the tissues to the lungs. Conversely, a rise in the partial pressure of CO_2 or a lower pH will cause offloading of oxygen from hemoglobin, which is known as the Bohr Effect.

Regulation of Respiration

Carbon dioxide is one of the mediators of local autoregulation of blood supply. If its concentration is high, the capillaries expand to allow a greater blood flow to that tissue.

Bicarbonate ions are crucial for regulating blood pH. A person's breathing rate influences the level of CO_2 in their blood. Breathing that is too slow or shallow causes respiratory acidosis, while breathing that is too rapid leads to hyperventilation, which can cause respiratory alkalosis.

Although the body requires oxygen for metabolism, low oxygen levels normally do not stimulate breathing. Rather, breathing is stimulated by higher carbon dioxide levels. As a result, breathing low-pressure air or a gas mixture with no oxygen at all (such as pure nitrogen) can lead to loss of consciousness without ever experiencing air hunger. This is especially perilous for high-altitude fighter pilots. It is also why flight attendants instruct passengers, in case of loss of cabin pressure, to apply the oxygen mask to themselves first before helping others; otherwise, one risks losing consciousness.

The respiratory centers try to maintain an arterial CO_2 pressure of 40 mm Hg. With intentional hyperventilation, the CO_2 content of arterial blood may be lowered to 10−20 mm Hg (the oxygen content of the blood is little affected), and the respiratory drive is diminished. This is why one can hold one's breath longer after hyperventilating than without hyperventilating. This carries the risk that unconsciousness may result before the need to breathe becomes overwhelming, which is why hyperventilation is particularly dangerous before free diving.

Methane

Methane is a chemical compound with the chemical formula CH_4 (one atom of carbon and four atoms of hydrogen). It is the simplest alkane and the main component of natural gas. The relative abundance of methane on Earth makes it an attractive fuel, though capturing and storing it poses challenges due to its gaseous state under normal conditions for temperature and pressure.

In its natural state, methane is found both below ground and under the sea floor. When it finds its way to the surface and the atmosphere, it is known as atmospheric methane. The Earth's atmospheric methane concentration has increased by about 150% since 1750, and it accounts for 20% of the total radiative forcing from all of the long-lived and globally mixed greenhouse gases (these gases don't include water vapor which is by far the largest component of the greenhouse effect). Methane breaks down in the atmosphere and creates $CH_3 \cdot$ with water vapor.

History

In November 1776, methane was first scientifically identified by Italian physicist Alessandro Volta in the marshes of Lake Maggiore straddling Italy and Switzerland. Volta was inspired to search for the substance after reading a paper written by Benjamin Franklin about "flammable air". Volta captured the gas rising from the marsh, and by 1778 had isolated the pure gas. He also demonstrated means to ignite the gas with an electric spark.

The name "methane" was coined in 1866 by the German chemist August Wilhelm von Hofmann.

Properties and Bonding

Methane is a tetrahedral molecule with four equivalent C–H bonds. Its electronic structure is described by four bonding molecular orbitals (MOs) resulting from the overlap of the valence orbitals on C and H. The lowest energy MO is the result of the overlap of the 2s orbital on carbon with the in-phase combination of the 1s orbitals on the four hydrogen atoms. Above this energy level is a triply degenerate set of MOs that involve overlap of the 2p orbitals on carbon with various linear combinations of the 1s orbitals on hydrogen. The resulting "three-over-one" bonding scheme is consistent with photoelectron spectroscopic measurements.

At room temperature and standard pressure, methane is a colorless, odorless gas. The familiar smell of natural gas as used in homes is achieved by the addition of an odorant, usually blends containing tert-butylthiol, as a safety measure. Methane has a boiling point of –161 °C (–257.8 °F) at a pressure of one atmosphere. As a gas it is flammable over a range of concentrations (4.4–17%) in air at standard pressure.

Solid methane exists in several modifications. Presently nine are known. Cooling methane at normal pressure results in the formation of methane I. This substance crystallizes in the cubic system (space group Fm3m). The positions of the hydrogen atoms are not fixed in methane I, i.e. methane molecules may rotate freely. Therefore, it is a plastic crystal.

Chemical Reactions

The primary chemical reactions of methane are combustion, steam reforming to syngas, and halogenation. In general, methane reactions are difficult to control. Partial oxidation to methanol, for example, is challenging because the reaction typically progresses all the way to carbon dioxide and water even with an insufficient supply of oxygen. The enzyme methane monooxygenase produces methanol from methane, but cannot be used for industrial-scale reactions.

Acid-base Reactions

Like other hydrocarbons, methane is a very weak acid. Its pKa in DMSO is estimated to be 56. It cannot be deprotonated in solution, but the conjugate base with methyllithium is known.

A variety of positive ions derived from methane have been observed, mostly as unstable species in low-pressure gas mixtures. These include methenium or methyl cation CH_3^+, methane cation CH_4^+, and methanium or protonated methane CH_5^+. Some of these have been detected in outer space. Methanium can also be produced as diluted solutions from methane with superacids. Cations with higher charge, such as CH_6^{2+} and CH_7^{3+}, have been studied theoretically and conjectured to be stable.

Despite the strength of its C–H bonds, there is intense interest in catalysts that facilitate C–H bond activation in methane (and other lower numbered alkanes).

Combustion

Methane's heat of combustion is 55.5 MJ/kg. Combustion of methane is a multiple step reaction. The following equations are part of the process, with the net result being:

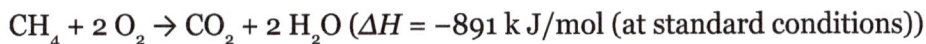

$$CH_4 + 2\,O_2 \rightarrow CO_2 + 2\,H_2O \quad (\Delta H = -891 \text{ k J/mol (at standard conditions))}$$

1. $CH_4 + M^* \rightarrow CH_3 + H + M$
2. $CH_4 + O_2 \rightarrow CH_3 + HO_2$
3. $CH_4 + HO_2 \rightarrow CH_3 + 2\,OH$
4. $CH_4 + OH \rightarrow CH_3 + H_2O$
5. $O_2 + H \rightarrow O + OH$
6. $CH_4 + O \rightarrow CH_3 + OH$
7. $CH_3 + O_2 \rightarrow CH_2O + OH$
8. $CH_2O + O \rightarrow CHO + OH$
9. $CH_2O + OH \rightarrow CHO + H_2O$
10. $CH_2O + H \rightarrow CHO + H_2$
11. $CHO + O \rightarrow CO + OH$
12. $CHO + OH \rightarrow CO + H_2O$
13. $CHO + H \rightarrow CO + H_2$
14. $H_2 + O \rightarrow H + OH$
15. $H_2 + OH \rightarrow H + H_2O$

16. $CO + OH \rightarrow CO_2 + H$

17. $H + OH + M \rightarrow H_2O + M^*$

18. $H + H + M \rightarrow H_2 + M^*$

19. $H + O_2 + M \rightarrow HO_2 + M^*$

The species M^* signifies an energetic third body, from which energy is transferred during a molecular collision. Formaldehyde (HCHO or H2CO) is an early intermediate (reaction 7). Oxidation of formaldehyde gives the formyl radical (HCO; reactions 8–10), which then give carbon monoxide (CO) (reactions 11, 12 & 13). Any resulting H_2 oxidizes to H_2O or other intermediates (reaction 14, 15). Finally, the CO oxidizes, forming CO_2 (reaction 16). In the final stages (reactions 17–19), energy is transferred back to other third bodies. The overall speed of reaction is a function of the concentration of the various entities during the combustion process. The higher the temperature, the greater the concentration of radical species and the more rapid the combustion process.

Reactions with Halogens

Given appropriate conditions, methane reacts with halogens as follows:

$X_2 + UV \rightarrow 2\,X\bullet$

$X\bullet + CH_4 \rightarrow HX + CH_3\bullet$

$CH_3\bullet + X_2 \rightarrow CH_3X + X\bullet$

where X is a halogen: fluorine (F), chlorine (Cl), bromine (Br), or iodine (I). This mechanism for this process is called free radical halogenation. It is initiated with UV light or some other radical initiator. A chlorine atom is generated from elemental chlorine, which abstracts a hydrogen atom from methane, resulting in the formation of hydrogen chloride. The resulting methyl radical, $CH_3\bullet$, can combine with another chlorine molecule to give methyl chloride (CH_3Cl) and a chlorine atom. This chlorine atom can then react with another methane (or methyl chloride) molecule, repeating the chlorination cycle. Similar reactions can produce dichloromethane (CH_2Cl_2), chloroform ($CHCl_3$), and, ultimately, carbon tetrachloride (CCl_4), depending upon reaction conditions and the chlorine to methane ratio.

Uses

Methane is used in industrial chemical processes and may be transported as a refrigerated liquid (liquefied natural gas, or LNG). While leaks from a refrigerated liquid container are initially heavier than air due to the increased density of the cold gas, the gas at ambient temperature is lighter than air. Gas pipelines distribute large amounts of natural gas, of which methane is the principal component.

Fuel

Methane is used as a fuel for ovens, homes, water heaters, kilns, automobiles, turbines, and other things. It combusts with oxygen to create fire.

Natural Gas

Methane is important for electrical generation by burning it as a fuel in a gas turbine or steam generator. Compared to other hydrocarbon fuels, methane produces less carbon dioxide for each unit of heat released. At about 891 kJ/mol, methane's heat of combustion is lower than any other hydrocarbon but the ratio of the heat of combustion (891 kJ/mol) to the molecular mass (16.0 g/mol, of which 12.0 g/mol is carbon) shows that methane, being the simplest hydrocarbon, produces more heat per mass unit (55.7 kJ/g) than other complex hydrocarbons. In many cities, methane is piped into homes for domestic heating and cooking. In this context it is usually known as natural gas, which is considered to have an energy content of 39 megajoules per cubic meter, or 1,000 BTU per standard cubic foot.

Methane in the form of compressed natural gas is used as a vehicle fuel and is claimed to be more environmentally friendly than other fossil fuels such as gasoline/petrol and diesel. Research into adsorption methods of methane storage for use as an automotive fuel has been conducted.

Liquefied Natural Gas

Liquefied natural gas (LNG) is natural gas (predominantly methane, CH_4) that has been converted to liquid form for ease of storage or transport.

Liquefied natural gas takes up about 1/600th the volume of natural gas in the gaseous state. It is odorless, colorless, non-toxic and non-corrosive. Hazards include flammability after vaporization into a gaseous state, freezing, and asphyxia.

The liquefaction process involves removal of certain components, such as dust, acid gases, helium, water, and heavy hydrocarbons, which could cause difficulty downstream. The natural gas is then condensed into a liquid at close to atmospheric pressure (maximum transport pressure set at around 25 kPa or 3.6 psi) by cooling it to approximately −162 °C (−260 °F).

LNG achieves a higher reduction in volume than compressed natural gas (CNG) so that the energy density of LNG is 2.4 times greater than that of CNG or 60% that of diesel fuel. This makes LNG cost efficient to transport over long distances where pipelines do not exist. Specially designed cryogenic sea vessels (LNG carriers) or cryogenic road tankers are used for its transport.

LNG, when it is not highly refined for special uses, is principally used for transporting natural gas to markets, where it is regasified and distributed as pipeline natural gas. It is also beginning to be used in LNG-fueled road vehicles. For example, trucks in commercial operation have been achieving payback periods of approximately four years on the higher initial investment required in LNG equipment on the trucks and LNG infrastructure to support fueling. However, it remains more common to design vehicles to use compressed natural gas. As of 2002, the relatively higher cost of LNG production and the need to store LNG in more expensive cryogenic tanks had slowed widespread commercial use.

Liquid Methane Rocket Fuel

In a highly refined form, liquid methane is used as a rocket fuel.

Though methane has been investigated for decades, no production methane engines have yet been used on orbital spaceflights. This promises to change as liquid methane has recently been selected for the active development of a variety of bipropellant rocket engines.

Since the 1990s, a number of Russian rockets using liquid methane have been proposed. One 1990s Russian engine proposal was the RD-192, a methane/LOX variant of the RD-191.

In 2005, US companies, Orbitech and XCOR Aerospace, developed a demonstration liquid oxygen/liquid methane rocket engine and a larger 7,500 pounds-force (33 kN)-thrust engine in 2007 for potential use as the CEV lunar return engine, before the CEV program was later cancelled.

More recently the American private space company SpaceX announced in 2012 an initiative to develop liquid methane rocket engines, including initially, the very large Raptor rocket engine. Raptor is being designed to produce 4.4 meganewtons (1,000,000 lbf) of thrust with a vacuum specific impulse (I_{sp}) of 363 seconds and a sea-level I_{sp} of 321 seconds, and began component-level testing in 2014. In February 2014, the Raptor engine design was shown to be of the highly efficient and theoretically more reliable full-flow staged combustion cycle type, where both propellant streams—oxidizer and fuel—are completely in the gas phase before they enter the combustion chamber. Prior to 2014, only two full-flow rocket engines had ever progressed sufficiently to be tested on test stands, but neither engine completed development or flew on a flight vehicle.

In October 2013, the China Aerospace Science and Technology Corporation, a state-owned contractor for the Chinese space program, announced that it had completed a first ignition test on a new LOX methane rocket engine. No engine size was provided.

In September 2014, another American private space company—Blue Origin— publicly announced that they were into their third year of development work on a large methane rocket engine. The new engine, the *Blue Engine 4*, or BE-4, has been designed to produce 2,400 kilonewtons (550,000 lbf) of thrust. While initially planned to be used exclusively on a Blue Origin proprietary launch vehicle, it will now be used on a new United Launch Alliance (ULA) engine on an new launch vehicle that is a successor to the Atlas V. ULA indicated in 2014 that they will make the maiden flight of the new launch vehicle no earlier than 2019.

One advantage of methane is that it is abundant in many parts of the solar system and it could potentially be harvested on the surface of another solar-system body (in particular, using methane production from local materials found on Mars or Titan), providing fuel for a return journey.

By 2013, NASA's Project Morpheus had developed a small restartable LOX methane rocket engine with 5,000 pounds-force (22 kN) thrust and a specific impulse of 321 seconds suitable for inspace applications including landers. Small LOX methane thrusters 5–15 pounds-force (22–67 N) were also developed suitable for use in a Reaction Control System (RCS).

SpaceNews is reporting in early 2015 that the French space agency CNES is working with Germany and a few other governments and will propose a LOX/methane engine on a reusable launch vehicle by mid-2015, with flight testing unlikely before approximately 2026.

Chemical Feedstock

Although there is great interest in converting methane into useful or more easily liquefied compounds, the only practical processes are relatively unselective. In the chemical industry, methane is converted to synthesis gas, a mixture of carbon monoxide and hydrogen, by steam reforming. This endergonic process (requiring energy) utilizes nickel catalysts and requires high temperatures, around 700–1100 °C:

$$CH_4 + H_2O \rightarrow CO + 3\ H_2$$

Related chemistries are exploited in the Haber-Bosch Synthesis of ammonia from air, which is reduced with natural gas to a mixture of carbon dioxide, water, and ammonia.

Methane is also subjected to free-radical chlorination in the production of chloromethanes, although methanol is a more typical precursor.

Production

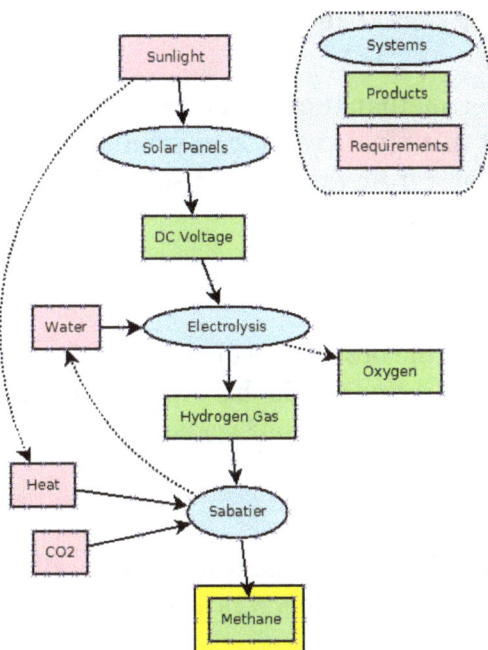

Biological Routes

Naturally occurring methane is mainly produced by the process of methanogenesis. This multistep process is used by microorganisms as an energy source. The net reaction is:

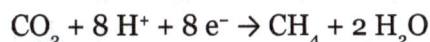

$$CO_2 + 8\ H^+ + 8\ e^- \rightarrow CH_4 + 2\ H_2O$$

The final step in the process is catalyzed by the enzyme Coenzyme-B sulfoethylthiotransferase. Methanogenesis is a form of anaerobic respiration used by organisms that occupy landfill, ruminants (e.g., cattle), and the guts of termites.

It is uncertain if plants are a source of methane emissions.

Power to Gas

Power to gas is a technology which converts electrical power to a gas fuel. The method is used to convert carbon dioxide and water to methane, using electrolysis and the Sabatier reaction. Excess and off-peak power generated by wind generators and solar arrays could theoretically be used for load balancing in the energy grid.

Industrial Routes

Methane can be produced by hydrogenating carbon dioxide through the Sabatier process. Methane is also a side product of the hydrogenation of carbon monoxide in the Fischer-Tropsch process. This technology is practiced on a large scale to produce longer chain molecules than methane.

Natural gas is so abundant that the intentional production of methane is relatively rare. The only large scale facility of this kind is the Great Plains Synfuels plant, started in 1984 in Beulah, North Dakota as a way to develop abundant local resources of low grade lignite, a resource which is otherwise very hard to transport for its weight, ash content, low calorific value and propensity to spontaneous combustion during storage and transport.

An adaptation of the Sabatier methanation reaction may be used via a mixed catalyst bed and a reverse water gas shift in a single reactor to produce methane from the raw materials available on Mars, utilizing water from the Martian subsoil and carbon dioxide in the Martian atmosphere.

Laboratory Synthesis

Methane can also be produced by the destructive distillation of acetic acid in the presence of soda lime or similar. Acetic acid is decarboxylated in this process. Methane can also be prepared by reaction of aluminium carbide with water or strong acids.

Serpentinization

Methane could also be produced by a non-biological process called *serpentinization* involving water, carbon dioxide, and the mineral olivine, which is known to be common on Mars.

Occurrence

Methane was discovered and isolated by Alessandro Volta between 1776 and 1778 when studying marsh gas from Lake Maggiore. It is the major component of natural gas, about 87% by volume. The major source of methane is extraction from geological deposits known as natural gas fields, with coal seam gas extraction becoming a major source (see Coal bed methane extraction, a method for extracting methane from a coal deposit, while enhanced coal bed methane recovery is a method of recovering methane from non-mineable coal seams). It is associated with other hydrocarbon fuels, and sometimes accompanied by helium and nitrogen. Methane is produced at shallow levels (low pressure) by anaerobic decay of organic matter and reworked methane from deep under the Earth's surface. In general, the sediments that generate natural gas are buried deeper and at higher temperatures than those that contain oil.

Methane is generally transported in bulk by pipeline in its natural gas form, or LNG carriers in its liquefied form; few countries transport it by truck.

Alternative Sources

Testing Australian sheep for exhaled methane production (2001), CSIRO

Apart from gas fields, an alternative method of obtaining methane is via biogas generated by the fermentation of organic matter including manure, wastewater sludge, municipal solid waste (including landfills), or any other biodegradable feedstock, under anaerobic conditions. Rice fields also generate large amounts of methane during plant growth. Methane hydrates/clathrates (ice-like combinations of methane and water on the sea floor, found in vast quantities) are a potential future source of methane. Cattle belch methane accounts for 16% of the world's annual methane emissions to the atmosphere. One study reported that the livestock sector in general (primarily cattle, chickens, and pigs) produces 37% of all human-induced methane. Early research has found a number of medical treatments and dietary adjustments that help slightly limit the production of methane in ruminants. A 2009 study found that at a conservative estimate, at least 51% of global greenhouse gas emissions were attributable to the life cycle and supply chain of livestock products, meaning all meat, dairy, and by-products, and their transportation. More recently, a 2013 study estimated that livestock accounted for 44 percent of human-induced methane and 14.5 percent of human-induced greenhouse gas emissions. Many efforts are underway to reduce livestock methane production and trap the gas to use as energy.

Paleoclimatology research published in *Current Biology* suggests that flatulence from dinosaurs may have warmed the Earth.

Atmospheric Methane

Methane is created near the Earth's surface, primarily by microorganisms by the process of methanogenesis. It is carried into the stratosphere by rising air in the tropics. Uncontrolled build-up of methane in the atmosphere is naturally checked – although human influence can upset this natural regulation – by methane's reaction with hydroxyl radicals formed from singlet oxygen atoms and with water vapor. It has a net lifetime of about 10 years, and is primarily removed by conversion to carbon dioxide and water.

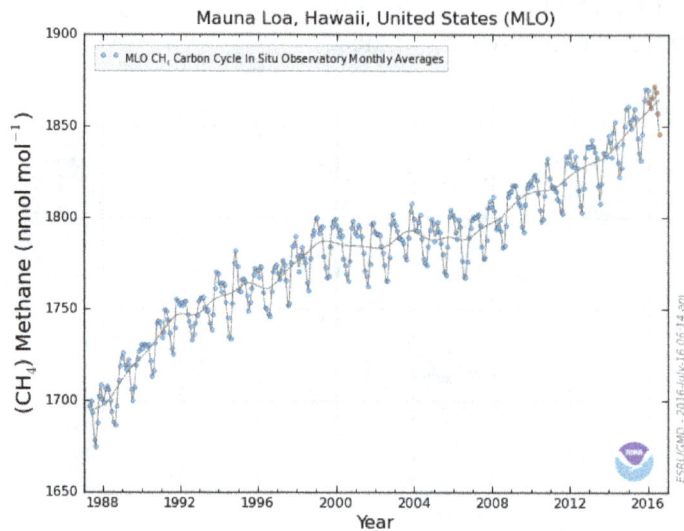

Methane concentrations up to December 2015 (Mauna Loa)

In addition, there is a large (but unknown) amount of methane in methane clathrates in the ocean floors as well as the Earth's crust.

In 2010, methane levels in the Arctic were measured at 1850 nmol/mol, a level over twice as high as at any time in the 400,000 years prior to the industrial revolution. Historically, methane concentrations in the world's atmosphere have ranged between 300 and 400 nmol/mol during glacial periods commonly known as ice ages, and between 600 and 700 nmol/mol during the warm interglacial periods. Recent research suggests that the Earth's oceans are a potentially important new source of Arctic methane.

Methane is an important greenhouse gas with a global warming potential of 34 compared to CO_2 over a 100-year period, and 72 over a 20-year period.

The Earth's atmospheric methane concentration has increased by about 150% since 1750, and it accounts for 20% of the total radiative forcing from all of the long-lived and globally mixed greenhouse gases (these gases don't include water vapor which is by far the largest component of the greenhouse effect).

Clathrates

Methane is essentially insoluble in water, but it can be trapped in ice forming a similar solid. Significant deposits of methane clathrate have been found under sediments on the ocean floors of Earth at large depths.

Arctic methane release from permafrost and methane clathrates is an expected consequence and further cause of global warming.

Anaerobic Oxidation of Methane

There is a group of bacteria that drive methane oxidation with nitrite as the oxidant, the anaerobic oxidation of methane.

Safety

Methane is nontoxic, yet it is extremely flammable and may form explosive mixtures with air. Methane is violently reactive with oxidizers, halogen, and some halogen-containing compounds. Methane is also an asphyxiant and may displace oxygen in an enclosed space. Asphyxia may result if the oxygen concentration is reduced to below about 16% by displacement, as most people can tolerate a reduction from 21% to 16% without ill effects. The concentration of methane at which asphyxiation risk becomes significant is much higher than the 5–15% concentration in a flammable or explosive mixture. Methane off-gas can penetrate the interiors of buildings near landfills and expose occupants to significant levels of methane. Some buildings have specially engineered recovery systems below their basements to actively capture this gas and vent it away from the building.

Methane gas explosions are responsible for many deadly mining disasters. A methane gas explosion was the cause of the Upper Big Branch coal mine disaster in West Virginia on April 5, 2010, killing 25.

Extraterrestrial Methane

Methane has been detected or is believed to exist on all planets of the solar system and most of the larger moons. With the possible exceptions of Mars and Titan, it is believed to have come from abiotic processes.

- Mercury – the tenuous atmosphere contains trace amounts of methane.

- Venus – the atmosphere contains a large amount of methane from 60 km (37 mi) to the surface according to data collected by the Pioneer Venus Large Probe Neutral Mass Spectrometer

- Moon – traces are outgassed from the surface

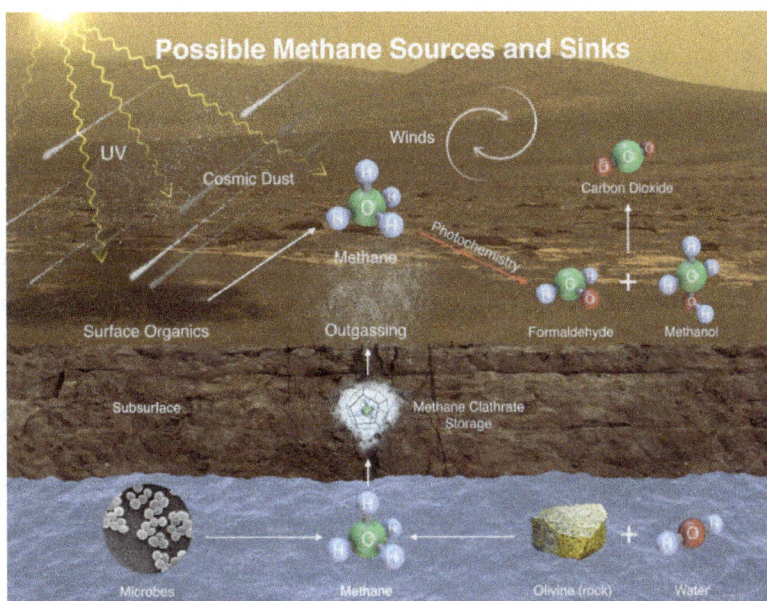

Methane (CH_4) on Mars – potential sources and sinks.

- Mars – the Martian atmosphere contains 10 nmol/mol methane. The source of methane on Mars has not been determined. Recent research suggests that methane may come from volcanoes, fault lines, or methanogens, that it may be a byproduct of electrical discharges from dust devils and dust storms, or that it may be the result of UV radiation. In January 2009, NASA scientists announced that they had discovered that the planet often vents methane into the atmosphere in specific areas, leading some to speculate this may be a sign of biological activity below the surface. Studies of a Weather Research and Forecasting model for Mars (MarsWRF) and related Mars general circulation model (MGCM) suggests that methane plume sources may be located within tens of kilometers, which is within the roving capabilities of future Mars rovers. The Curiosity rover, which landed on Mars in August 2012, can distinguish between different isotopologues of methane; but even if the mission determines that microscopic Martian life is the source of the methane, it probably resides far below the surface, beyond the rover's reach. Curiosity's Sample Analysis at Mars (SAM) instrument is capable of tracking the presence of methane over time to determine if it is constant, variable, seasonal, or random, providing further clues about its source. The first measurements with the Tunable Laser Spectrometer (TLS) indicated that there is less than 5 ppb of methane at the landing site.

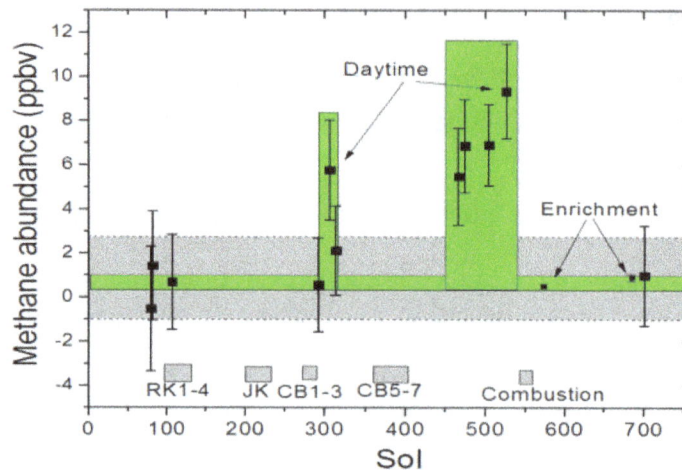

Methane measurements in the atmosphere of Mars by the *Curiosity* rover.

The Mars Trace Gas Mission orbiter planned for launch in 2016 would further study Mars' methane and its decomposition products such as formaldehyde and methanol. Alternatively, these compounds may instead be replenished by volcanic or other geological means, such as serpentinization. On July 19, 2013, NASA scientists reported finding "not much methane" (i.e., "an upper limit of 2.7 parts per billion of methane") around the Gale Crater where the Curiosity rover landed in August 2012. On September 19, 2013, from further measurements by Curiosity, NASA scientists reported no detection of atmospheric methane with a value of 0.18±0.67 ppbv corresponding to an upper limit of only 1.3 ppbv (95% confidence limit), and as a result, concluded that the probability of current methanogenic microbial activity on Mars is reduced. On 16 December 2014, NASA reported the *Curiosity* rover detected a "tenfold spike", likely localized, in the amount of methane in the Martian atmosphere. Sample measurements taken "a dozen times over 20 months" showed increases in late 2013 and early 2014, averaging "7 parts of methane per billion in the atmosphere." Before and after that, readings averaged around one-tenth that level.

- Saturn – the atmosphere contains 4500 ± 2000 ppm methane

- Enceladus – the atmosphere contains 1.7% methane

- Iapetus

- Titan – the atmosphere contains 1.6% methane and thousands of methane lakes have been detected on the surface. In the upper atmosphere, methane is converted into more complex molecules including acetylene, a process that also produces molecular hydrogen. There is evidence that acetylene and hydrogen are recycled into methane near the surface. This suggests the presence either of an exotic catalyst, possibly an unknown form of methanogenic life. Methane showers, probably prompted by changing seasons, have also been observed. On October 24, 2014, methane was found in polar clouds on Titan.

Polar clouds, made of methane, on Titan (left) compared with polar clouds on Earth (right).

- Uranus – the atmosphere contains 2.3% methane

  - Ariel – methane is believed to be a constituent of Ariel's surface ice

  - Miranda

  - Oberon – about 20% of Oberon's surface ice is composed of methane-related carbon/nitrogen compounds

  - Titania – about 20% of Titania's surface ice is composed of methane-related organic compounds

  - Umbriel – methane is a constituent of Umbriel's surface ice

- Neptune – the atmosphere contains 1.5 ± 0.5% methane

  - Triton – Triton has a tenuous nitrogen atmosphere with small amounts of methane near the surface.

- Pluto – spectroscopic analysis of Pluto's surface reveals it to contain traces of methane

  - Charon – methane is believed present on Charon, but it is not completely confirmed

- Eris – infrared light from the object revealed the presence of methane ice

- Halley's Comet

- Comet Hyakutake – terrestrial observations found ethane and methane in the comet

- Extrasolar planets – methane was detected on extrasolar planet HD 189733b; this is the first detection of an organic compound on a planet outside the solar system. Its origin is unknown, since the planet's high temperature (700 °C) would normally favor the formation of carbon monoxide instead. Research indicates that meteoroids slamming against exoplanet atmospheres could add hydrocarbon gases such as methane, making the exoplanets look as though they are inhabited by life, even if they are not.

- Interstellar clouds

- The atmospheres of M-type stars.

Ozone

Ozone (systematically named $1\lambda^1,3\lambda^1$-trioxidane and catena-trioxygen), or trioxygen, is an inorganic molecule with the chemical formula O_3. It is a pale blue gas with a distinctively pungent smell. It is an allotrope of oxygen that is much less stable than the diatomic allotrope O_2, breaking down in the lower atmosphere to normal dioxygen. Ozone is formed from dioxygen by the action of ultraviolet light and also atmospheric electrical discharges, and is present in low concentrations throughout the Earth's atmosphere (stratosphere). In total, ozone makes up only 0.6 ppm of the atmosphere.

Ozone's odour is sharp, reminiscent of chlorine, and detectable by many people at concentrations of as little as 10 ppb in air. Ozone's O_3 structure was determined in 1865. The molecule was later proven to have a bent structure and to be diamagnetic. In standard conditions, ozone is a pale blue gas that condenses at progressively cryogenic temperatures to a dark blue liquid and finally a violet-black solid. Ozone's instability with regard to more common dioxygen is such that both concentrated gas and liquid ozone may decompose explosively at elevated temperatures or fast warming to the boiling point. It is therefore used commercially only in low concentrations.

Ozone is a powerful oxidant (far more so than dioxygen) and has many industrial and consumer applications related to oxidation. This same high oxidising potential, however, causes ozone to damage mucous and respiratory tissues in animals, and also tissues in plants, above concentrations of about 100 ppb. This makes ozone a potent respiratory hazard and pollutant near ground level. However, the ozone layer (a portion of the stratosphere with a bigger concentration of ozone, from two to eight ppm) is beneficial, preventing damaging ultraviolet light from reaching the Earth's surface, to the benefit of both plants and animals.

Nomenclature

The trivial name *ozone* is the most commonly used and preferred IUPAC name. The systematic names $1\lambda^1,3\lambda^1$-*trioxidane* and *catena-trioxygen*, valid IUPAC names, are constructed according to the substitutive and additive nomenclatures, respectively. The name *ozone* derives from *ozein*, the Greek verb for smell, referring to ozone's distinctive smell.

In appropriate contexts, ozone can be viewed as trioxidane with two hydrogen atoms removed, and as such, *trioxidanylidene* may be used as a context-specific systematic name, according to substitutive nomenclature. By default, these names pay no regard to the radicality of the ozone molecule. In even more specific context, this can also name the non-radical singlet ground state, whereas the diradical state is named *trioxidanediyl*.

Trioxidanediyl (or *ozonide*) is used, non-systematically, to refer to the substituent group (-OOO-). Care should be taken to avoid confusing the name of the group for the context-specific name for ozone given above.

History

Christian Friedrich Schönbein (18 October 1799 – 29 August 1868)

A prototype ozonometer built by John Smyth in 1865

In 1785, the Dutch chemist Martinus van Marum was conducting experiments involving electrical sparking above water when he noticed an unusual smell, which he attributed to the electrical reactions, failing to realize that he had in fact created ozone. A half century later, Christian Friedrich Schönbein noticed the same pungent odour and recognized it as the smell often following a bolt of lightning. In 1839, he succeeded in isolating the gaseous chemical and named it "ozone", from the Greek word *ozein* meaning "to smell". For this reason, Schönbein is generally credited with the discovery of ozone. The formula for ozone, O_3, was not determined until 1865 by Jacques-Louis Soret and confirmed by Schönbein in 1867.

For much of the second half of the nineteenth century and well into the twentieth, ozone was considered a healthy component of the environment by naturalists and health-seekers. Beaumont, California had as its official slogan "Beaumont: Zone of Ozone", as evidenced on postcards and Chamber of Commerce letterhead. Naturalists working outdoors often considered the higher elevations beneficial because of their ozone content. "There is quite a different atmosphere [at higher elevation] with enough ozone to sustain the necessary energy [to work]", wrote naturalist Henry Henshaw, working in Hawaii. Seaside air was considered to be healthy because of its believed ozone content; but the smell giving rise to this belief is in fact that of halogenated seaweed metabolites.

In fact, even Benjamin Franklin believed that the presence of cholera was connected with the deficiency or lack of ozone in the atmosphere, a sentiment shared by the British Science Association (then known simply as the British Association).

Physical Properties

Ozone is colourless or slightly bluish gas (blue when liquefied), slightly soluble in water and much more soluble in inert non-polar solvents such as carbon tetrachloride or fluorocarbons, where it forms a blue solution. At 161 K (−112 °C; −170 °F), it condenses to form a dark blue liquid. It is dangerous to allow this liquid to warm to its boiling point, because both concentrated gaseous ozone and liquid ozone can detonate. At temperatures below 80 K (−193.2 °C; −315.7 °F), it forms a violet-black solid.

Most people can detect about 0.01 μmol/mol of ozone in air where it has a very specific sharp odour somewhat resembling chlorine bleach. Exposure of 0.1 to 1 μmol/mol produces headaches, burning eyes and irritation to the respiratory passages. Even low concentrations of ozone in air are very destructive to organic materials such as latex, plastics and animal lung tissue.

Ozone is diamagnetic, which means that its electrons are all paired. In contrast, O_2 is paramagnetic, containing two unpaired electrons.

Structure

According to experimental evidence from microwave spectroscopy, ozone is a bent molecule, with C_{2v} symmetry (similar to the water molecule). The O − O distances are 127.2 pm (1.272 Å). The O − O − O angle is 116.78°. The central atom is sp^2 hybridized with one lone pair. Ozone is a polar molecule with a dipole moment of 0.53 D. The bonding can be expressed as a resonance hybrid with a single bond on one side and double bond on the other producing an overall bond order of 1.5 for each side.

Reactions

Ozone is a powerful oxidizing agent, far stronger than O_2. It is also unstable at high concentrations, decaying to ordinary diatomic oxygen. It has a varying half-life length, depending upon atmospheric conditions (temperature, humidity, and air movement). In a sealed chamber with a fan that moves the gas, ozone has a half-life of approximately a day at room temperature. Some unverified claims imply that ozone can have a half life as short as a half an hour under atmospheric conditions.

$$2O_3 \rightarrow 3O_2$$

This reaction proceeds more rapidly with increasing temperature and increased pressure. Deflagration of ozone can be triggered by a spark, and can occur in ozone concentrations of 10 wt% or higher.

With Metals

Ozone will oxidise most metals (except gold, platinum, and iridium) to oxides of the metals in their highest oxidation state. For example:

$$Cu + O_3 \rightarrow CuO + O_2$$

With Nitrogen and Carbon Compounds

Ozone also oxidizes nitric oxide to nitrogen dioxide:

$$NO + O_3 \rightarrow NO_2 + O_2$$

This reaction is accompanied by chemiluminescence. The NO2 can be further oxidized:

$$NO_2 + O_3 \rightarrow NO_3 + O_2$$

The NO_3 formed can react with NO_2 to form N_2O_5.

Solid nitronium perchlorate can be made from NO_2, ClO_2, and O_3 gases:

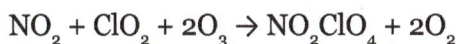

$$NO_2 + ClO_2 + 2O_3 \rightarrow NO_2ClO_4 + 2O_2$$

Ozone does not react with ammonium salts, but it oxidizes ammonia to ammonium nitrate:

$$2NH_3 + 4O_3 \rightarrow NH_4NO_3 + 4O_2 + H_2O$$

Ozone reacts with carbon to form carbon dioxide, even at room temperature:

$$C + 2O_3 \rightarrow CO_2 + 2O_2$$

With Sulfur Compounds

Ozone oxidises sulfides to sulfates. For example, lead(II) sulfide is oxidised to lead(II) sulfate:

$$PbS + 4O_3 \rightarrow PbSO_4 + 4O_2$$

Sulfuric acid can be produced from ozone, water and either elemental sulfur or sulfur dioxide:

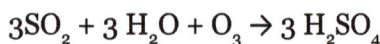

$$S + H_2O + O_3 \rightarrow H_2SO_4$$

$$3SO_2 + 3 H_2O + O_3 \rightarrow 3 H_2SO_4$$

In the gas phase, ozone reacts with hydrogen sulfide to form sulfur dioxide:

$$H_2S + O_3 \rightarrow SO_2 + H_2O$$

In an aqueous solution, however, two competing simultaneous reactions occur, one to produce elemental sulfur, and one to produce sulfuric acid:

$$H_2S + O_3 \rightarrow S + O_2 + H_2O$$

$$3 H_2S + 4 O_3 \rightarrow 3 H_2SO_4$$

With Alkenes and Alkynes

Alkenes can be oxidatively cleaved by ozone, in a process called ozonolysis, giving alcohols, aldehydes, ketones, and carboxylic acids, depending on the second step of the workup.

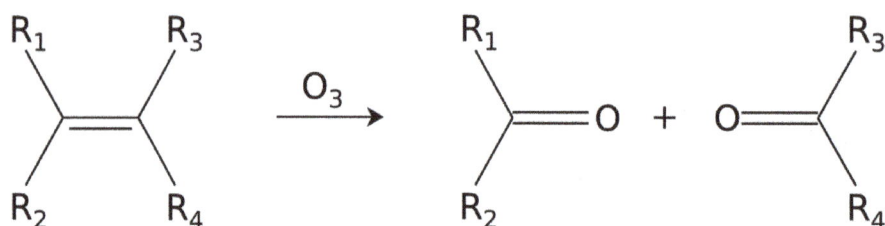

Usually ozonolysis is carried out in a solution of dichloromethane, at a temperature of −78°C. After a sequence of cleavage and rearrangement, an organic ozonide is formed. With reductive workup (e.g. zinc in acetic acid or dimethyl sulfide), ketones and aldehydes will be formed, with oxidative workup (e.g. aqueous or alcoholic hydrogen peroxide), carboxylic acids will be formed.

Other Substrates

All three atoms of ozone may also react, as in the reaction of tin(II) chloride with hydrochloric acid and ozone:

$$3SnCl_2 + 6HCl + O3 \rightarrow 3SnCl_4 + 3H_2O$$

Iodine perchlorate can be made by treating iodine dissolved in cold anhydrous perchloric acid with ozone:

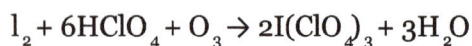

$$I_2 + 6HClO_4 + O_3 \rightarrow 2I(ClO_4)_3 + 3H_2O$$

Combustion

Ozone can be used for combustion reactions and combustible gases; ozone provides higher temperatures than burning in dioxygen (O_2). The following is a reaction for the combustion of carbon subnitride which can also cause higher temperatures:

$$3C_4N_2 + 4O_3 \rightarrow 12CO + 3N_2$$

Ozone can react at cryogenic temperatures. At 77 K (−196.2 °C; −321.1 °F), atomic hydrogen reacts with liquid ozone to form a hydrogen superoxide radical, which dimerizes:

$$H + O_3 \rightarrow HO_2 + O$$

$$2HO_2 \rightarrow H_2O_4$$

Reduction to Ozonides

Reduction of ozone gives the ozonide anion, O_3^-. Derivatives of this anion are explosive and must be stored at cryogenic temperatures. Ozonides for all the alkali metals are known. KO_3, RbO_3, and CsO_3 can be prepared from their respective superoxides:

$$KO_2 + O_3 \rightarrow KO_3 + O_2$$

Although KO_3 can be formed as above, it can also be formed from potassium hydroxide and ozone:

$$2KOH + 5O_3 \rightarrow 2KO_3 + 5O_2 + H_2O$$

NaO_3 and LiO_3 must be prepared by action of CsO_3 in liquid NH_3 on an ion exchange resin containing Na^+ or Li^+ ions:

$$CsO_3 + Na^+ \rightarrow Cs^+ + NaO_3$$

A solution of calcium in ammonia reacts with ozone to give to ammonium ozonide and not calcium ozonide:

$$3Ca + 10NH_3 + 6O3 \rightarrow Ca·6NH_3 + Ca(OH)_2 + Ca(NO_3)_2 + 2NH_4O_3 + 2O_2 + H_2$$

Applications

Ozone can be used to remove iron and manganese from water, forming a precipitate which can be filtered:

$$2Fe^{2+} + O_3 + 5H_2O \rightarrow 2Fe(OH)_3(s) + O_2 + 4H^+$$

$$2Mn^{2+} + 2O_3 + 4H_2O \rightarrow 2MnO(OH)_2(s) + 2O_2 + 4H^+$$

Ozone will also reduce dissolved hydrogen sulfide in water to sulfurous acid:

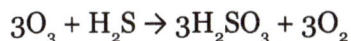

$$3O_3 + H_2S \rightarrow 3H_2SO_3 + 3O_2$$

These three reactions are central in the use of ozone based well water treatment.

Ozone will also detoxify cyanides by converting them to cyanates.

$$CN^- + O_3 \rightarrow CNO^- + O_2$$

Ozone will also completely decompose urea:

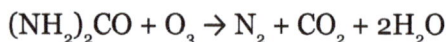

$$(NH_2)_2CO + O_3 \rightarrow N_2 + CO_2 + 2H_2O$$

Ozone Spectroscopy

Ozone is a bent triatomic molecule with three vibrational modes: the symmetric stretch (1103.157 cm^{-1}), bend (701.42 cm^{-1}) and antisymmetric stretch (1042.096 cm^{-1}). The symmetric stretch and bend are weak absorbers, but the antisymmetric stretch is strong and responsible for ozone being an important minor greenhouse gas. This IR band is also used to detect ambient and atmospheric ozone although UV based measurements are more common.

The electronic spectrum of ozone is quite complex. An overview can be seen at the MPI Mainz UV/ VIS Spectral Atlas of Gaseous Molecules of Atmospheric Interest.

All of the bands are dissociative, meaning that the molecule falls apart to $O + O_2$ after absorbing a photon. The most important absorption is the Hartley band, extending from slightly above 300 nm down to slightly above 200 nm. It is this band that is responsible for absorbing UV C in the stratosphere.

On the high wavelength side, the Hartley band transitions to the so-called Huggins band, which falls off rapidly till disappearing by ~360 nm. Above 400 nm, extending well out into the NIR, are the Chappius and Wulf bands. There, unstructured absorption bands are useful for detecting high ambient concentrations of ozone, but are so weak that they do not have much practical effect.

There are additional absorption bands in the far UV, which increase slowly from 200 nm down to reaching a maximum at ~120 nm.

Ozone in Earth's Atmosphere

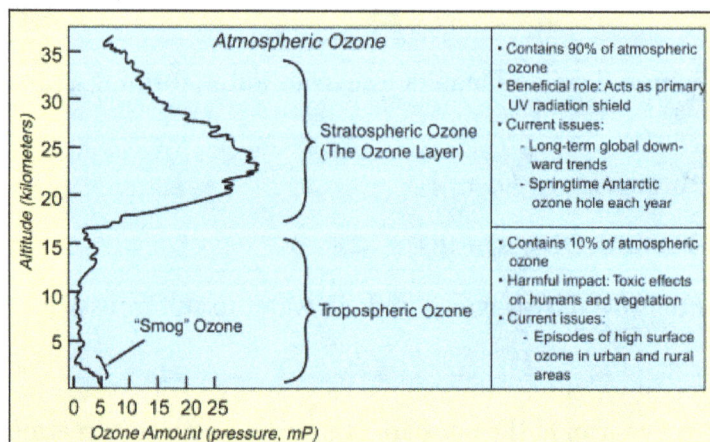

The distribution of atmospheric ozone in partial pressure as a function of altitude

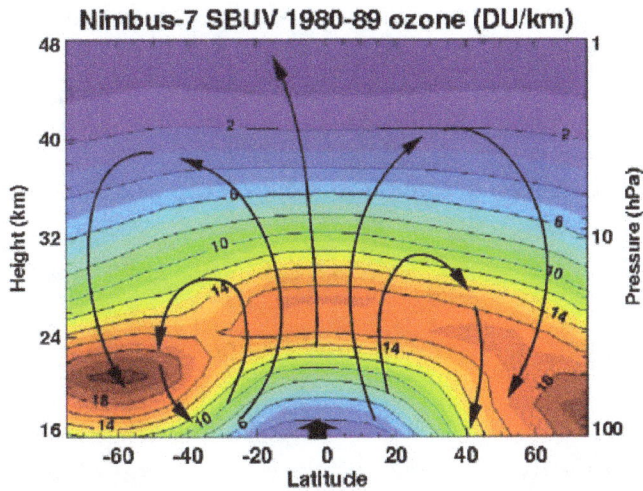

Concentration of ozone as measured by the Nimbus-7 satellite

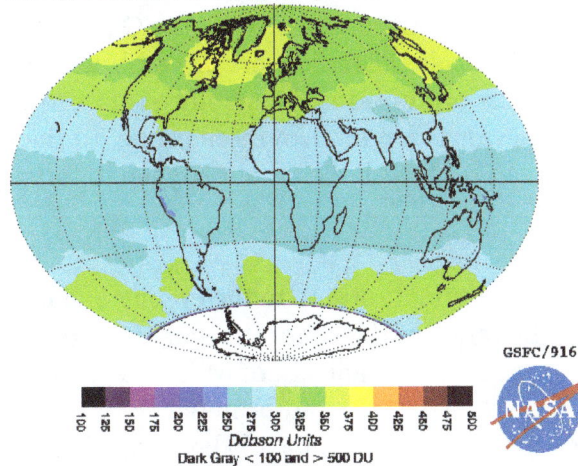

Total ozone concentration in June 2000 as measured by EP-TOMS satellite instrument

The standard way to express total ozone levels (the amount of ozone in a given vertical column) in the atmosphere is by using Dobson units. Point measurements are reported as mole fractions in nmol/mol (parts per billion, ppb) or as concentrations in $\mu g/m^3$. The study of ozone concentration in the atmosphere started in the 1920s.

Ozone Layer

Location and Production

The highest levels of ozone in the atmosphere are in the stratosphere, in a region also known as the ozone layer between about 10 km and 50 km above the surface (or between about 6 and 31 miles). However, even in this "layer", the ozone concentrations are only two to eight parts per million, so most of the oxygen there remains of the dioxygen type.

Ozone in the stratosphere is mostly produced from short-wave ultraviolet rays between 240 and 160 nm. Oxygen starts to absorb weakly at 240 nm in the Herzberg bands, but most of the oxygen is dissociated by absorption in the strong Schumann–Runge bands between 200 and 160 nm where ozone does not absorb. While shorter wavelength light, extending to even the X-Ray limit, is energetic enough to dissociate molecular oxygen, there is relatively little of it, and, the strong solar emission at Lyman-alpha, 121 nm, falls at a point where molecular oxygen absorption is a minimum.

The process of ozone creation and destruction is called the Chapman cycle and starts with the photolysis of molecular oxygen

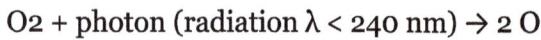

O_2 + photon (radiation $\lambda < 240$ nm) $\rightarrow$ 2 O

followed by reaction of the oxygen atom with another molecule of oxygen to form ozone.

$O + O_2 + M \rightarrow O_3 + M$

where "M" denotes the third body that carries off the excess energy of the reaction. The ozone molecule can then absorb a UVC photon and dissociate

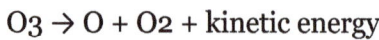

$O_3 \rightarrow O + O_2$ + kinetic energy

The excess kinetic energy heats the stratosphere when the O atoms and the molecular oxygen fly apart and collide with other molecules. This conversion of UV light into kinetic energy warms the stratosphere. The oxygen atoms produced in the photolysis of ozone then react back with other oxygen molecule as in the previous step to form more ozone. In the clear atmosphere, with only nitrogen and oxygen, ozone can react with the atomic oxygen to form two molecules of O_2

$O_3 + O \rightarrow 2 O_2$

An estimate of the rate of this termination step to the cycling of atomic oxygen back to ozone can be found simply by taking the ratios of the concentration of O_2 to O_3. The termination reaction is catalysed by the presence of certain free radicals, of which the most important are hydroxyl (OH), nitric oxide (NO) and atomic chlorine (Cl) and bromine (Br). In recent decades, the amount of ozone in the stratosphere has been declining, mostly because of emissions of chlorofluorocarbons (CFC) and similar chlorinated and brominated organic molecules, which have increased the concentration of ozone-depleting catalysts above the natural background.

Importance to Surface-dwelling Life on Earth

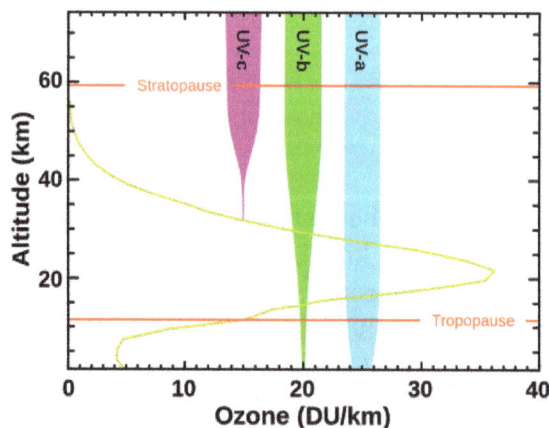

Levels of ozone at various altitudes and blocking of different bands of ultraviolet radiation. Essentially all UVC (100–280 nm) is blocked by dioxygen (at 100–200 nm) or by ozone (at 200–280 nm) in the atmosphere. The shorter portion of this band and even more energetic UV causes the formation of the ozone layer, when single oxygen atoms produced by UV photolysis of dioxygen (below 240 nm) react with more dioxygen. The ozone layer itself then blocks most, but not quite all, sunburn-producing UVB (280–315 nm). The band of UV closest to visible light, UVA (315–400 nm), is hardly affected by ozone, and most of it reaches the ground.

Ozone in the ozone layer filters out sunlight wavelengths from about 200 nm UV rays to 315 nm, with ozone peak absorption at about 250 nm. This ozone UV absorption is important to life, since it extends the absorption of UV by ordinary oxygen and nitrogen in air (which absorb all wavelengths < 200 nm) through the lower UV-C (200–280 nm) and the entire UV-B band (280–315 nm). The small unabsorbed part that remains of UV-B after passage through ozone causes sunburn in humans, and direct DNA damage in living tissues in both plants and animals. Ozone's effect on mid-range UV-B rays is illustrated by its effect on UV-B at 290 nm, which has a radiation intensity 350 million times as powerful at the top of the atmosphere as at the surface. Nevertheless, enough of UV-B radiation at similar frequency reaches the ground to cause some sunburn, and these same wavelengths are also among those responsible for the production of vitamin D in humans.

The ozone layer has little effect on the longer UV wavelengths called UV-A (315–400 nm), but this radiation does not cause sunburn or direct DNA damage, and while it probably does cause long-term skin damage in certain humans, it is not as dangerous to plants and to the health of surface-dwelling organisms on Earth in general.

Low Level Ozone

Low level ozone (or tropospheric ozone) is an atmospheric pollutant. It is not emitted directly by car engines or by industrial operations, but formed by the reaction of sunlight on air containing hydrocarbons and nitrogen oxides that react to form ozone directly at the source of the pollution or many kilometers down wind.

Ozone reacts directly with some hydrocarbons such as aldehydes and thus begins their removal from the air, but the products are themselves key components of smog. Ozone photolysis by UV light leads to production of the hydroxyl radical HO• and this plays a part in the removal of hydrocarbons from the air, but is also the first step in the creation of components of smog such as peroxyacyl nitrates, which can be powerful eye irritants. The atmospheric lifetime of tropospheric ozone is about 22 days; its main removal mechanisms are being deposited to the ground, the above-mentioned reaction giving HO•, and by reactions with OH and the peroxy radical HO_2•.

There is evidence of significant reduction in agricultural yields because of increased ground-level ozone and pollution which interferes with photosynthesis and stunts overall growth of some plant species. The United States Environmental Protection Agency is proposing a secondary regulation to reduce crop damage, in addition to the primary regulation designed for the protection of human health.

Certain examples of cities with elevated ozone readings are Houston, Texas, and Mexico City,

Mexico. Houston has a reading of around 41 nmol/mol, while Mexico City is far more hazardous, with a reading of about 125 nmol/mol.

Ozone Cracking

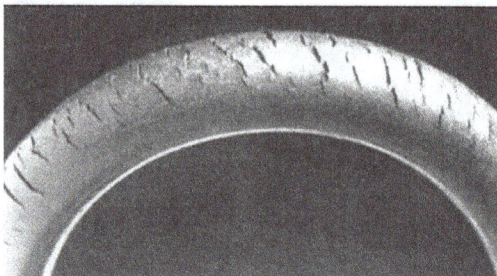

Ozone cracking in natural rubber tubing

Ozone gas attacks any polymer possessing olefinic or double bonds within its chain structure, such as natural rubber, nitrile rubber, and styrene-butadiene rubber. Products made using these polymers are especially susceptible to attack, which causes cracks to grow longer and deeper with time, the rate of crack growth depending on the load carried by the rubber component and the concentration of ozone in the atmosphere. Such materials can be protected by adding antiozonants, such as waxes, which bond to the surface to create a protective film or blend with the material and provide long term protection. Ozone cracking used to be a serious problem in car tires for example, but the problem is now seen only in very old tires. On the other hand, many critical products, like gaskets and O-rings, may be attacked by ozone produced within compressed air systems. Fuel lines made of reinforced rubber are also susceptible to attack, especially within the engine compartment, where some ozone is produced by electrical components. Storing rubber products in close proximity to a DC electric motor can accelerate ozone cracking. The commutator of the motor generates sparks which in turn produce ozone.

Ozone as a Greenhouse Gas

Although ozone was present at ground level before the Industrial Revolution, peak concentrations are now far higher than the pre-industrial levels, and even background concentrations well away from sources of pollution are substantially higher. Ozone acts as a greenhouse gas, absorbing some of the infrared energy emitted by the earth. Quantifying the greenhouse gas potency of ozone is difficult because it is not present in uniform concentrations across the globe. However, the most widely accepted scientific assessments relating to climate change (e.g. the Intergovernmental Panel on Climate Change Third Assessment Report) suggest that the radiative forcing of tropospheric ozone is about 25% that of carbon dioxide.

The annual global warming potential of tropospheric ozone is between 918–1022 tons carbon dioxide equivalent/tons tropospheric ozone. This means on a per-molecule basis, ozone in the troposphere has a radiative forcing effect roughly 1,000 times as strong as carbon dioxide. However, tropospheric ozone is a short-lived greenhouse gas, which decays in the atmosphere much more quickly than carbon dioxide. This means that over a 20-year span, the global warming potential of tropospheric ozone is much less, roughly 62 to 69 tons carbon dioxide equivalent / ton tropospheric ozone.

Because of its short-lived nature, tropospheric ozone does not have strong global effects, but has very strong radiative forcing effects on regional scales. In fact, there are regions of the world where tropospheric ozone has a radiative forcing up to 150% of carbon dioxide.

Health Effects

Ozone Air Pollution

Red Alder leaf, showing discolouration caused by ozone pollution

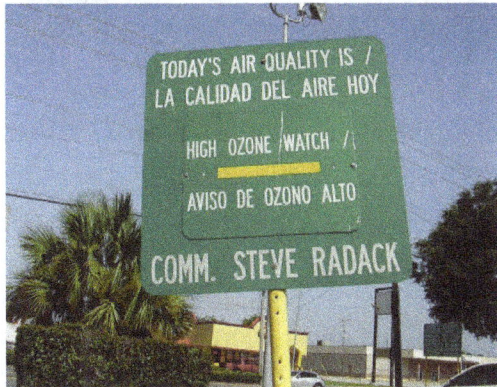

Signboard in Gulfton, Houston indicating an ozone watch

Ozone precursors are a group of pollutants, predominantly those emitted during the combustion of fossil fuels. Ground-level ozone pollution (tropospheric ozone) is created near the Earth's surface by the action of daylight UV rays on these precursors. The ozone at ground level is primarily from fossil fuel precursors, but methane is a natural precursor, and the very low natural background level of ozone at ground level is considered safe. This section examines the health impacts of fossil fuel burning, which raises ground level ozone far above background levels.

There is a great deal of evidence to show that ground level ozone can harm lung function and irritate the respiratory system. Exposure to ozone (and the pollutants that produce it) is linked to premature death, asthma, bronchitis, heart attack, and other cardiopulmonary problems.

Long-term exposure to ozone has been shown to increase risk of death from respiratory illness. A study of 450,000 people living in United States cities saw a significant correlation between ozone levels and respiratory illness over the 18-year follow-up period. The study revealed that people living in cities with high ozone levels, such as Houston or Los Angeles, had an over 30% increased risk of dying from lung disease.

Air quality guidelines such as those from the World Health Organization, the United States Environmental Protection Agency (EPA) and the European Union are based on detailed studies designed to identify the levels that can cause measurable ill health effects.

According to scientists with the US EPA, susceptible people can be adversely affected by ozone levels as low as 40 nmol/mol. In the EU, the current target value for ozone concentrations is 120 µg/m³ which is about 60 nmol/mol. This target applies to all member states in accordance with Directive 2008/50/EC. Ozone concentration is measured as a maximum daily mean of 8 hour averages and the target should not be exceeded on more than 25 calendar days per year, starting from January 2010. Whilst the directive requires in the future a strict compliance with 120 µg/m³ limit (i.e. mean ozone concentration not to be exceeded on any day of the year), there is no date set for this requirement and this is treated as a long-term objective.

In the USA, the Clean Air Act directs the EPA to set National Ambient Air Quality Standards for several pollutants, including ground-level ozone, and counties out of compliance with these standards are required to take steps to reduce their levels. In May 2008, under a court order, the EPA lowered its ozone standard from 80 nmol/mol to 75 nmol/mol. The move proved controversial, since the Agency's own scientists and advisory board had recommended lowering the standard to 60 nmol/mol. Many public health and environmental groups also supported the 60 nmol/mol standard, and the World Health Organization recommends 51 nmol/mol.

On January 7, 2010, the U.S. Environmental Protection Agency (EPA) announced proposed revisions to the National Ambient Air Quality Standard (NAAQS) for the pollutant ozone, the principal component of smog:

... EPA proposes that the level of the 8-hour primary standard, which was set at 0.075 µmol/mol in the 2008 final rule, should instead be set at a lower level within the range of 0.060 to 0.070 µmol/mol, to provide increased protection for children and other "at risk" populations against an array of O_3-related adverse health effects that range from decreased lung function and increased respiratory symptoms to serious indicators of respiratory morbidity including emergency department visits and hospital admissions for respiratory causes, and possibly cardiovascular-related morbidity as well as total non- accidental and cardiopulmonary mortality....

On October 26, 2015, the EPA published a final rule with an effective date of December 28, 2015 that revised the 8-hour primary NAAQS from 0.075 ppm to 0.070 ppm.

The EPA has developed an Air Quality Index (AQI) to help explain air pollution levels to the general public. Under the current standards, eight-hour average ozone mole fractions of 85 to 104 nmol/mol are described as "unhealthy for sensitive groups", 105 nmol/mol to 124 nmol/mol as "unhealthy", and 125 nmol/mol to 404 nmol/mol as "very unhealthy".

Ozone can also be present in indoor air pollution, partly as a result of electronic equipment such as photocopiers. A connection has also been known to exist between the increased pollen, fungal spores, and ozone caused by thunderstorms and hospital admissions of asthma sufferers.

In the Victorian era, one British folk myth held that the smell of the sea was caused by ozone. In fact, the characteristic "smell of the sea" is caused by dimethyl sulfide, a chemical generated by phytoplankton. Victorian British folk considered the resulting smell "bracing".

Heat Waves

Ozone production rises during heat waves, because plants absorb less ozone. It is estimated that curtailed ozone absorption by plants was responsible for the loss of 460 lives in the UK in the hot summer of 2006. A similar investigation to assess the joint effects of ozone and heat during the European heat waves in 2003, concluded that these appear to be additive.

Physiology

Ozone, along with reactive forms of oxygen such as superoxide, singlet oxygen, hydrogen peroxide, and hypochlorite ions, is naturally produced by white blood cells and other biological systems (such as the roots of marigolds) as a means of destroying foreign bodies. Ozone reacts directly with organic double bonds. Also, when ozone breaks down to dioxygen it gives rise to oxygen free radicals, which are highly reactive and capable of damaging many organic molecules. Moreover, it is believed that the powerful oxidizing properties of ozone may be a contributing factor of inflammation. The cause-and-effect relationship of how the ozone is created in the body and what it does is still under consideration and still subject to various interpretations, since other body chemical processes can trigger some of the same reactions. A team headed by Paul Wentworth Jr. of the Department of Chemistry at the Scripps Research Institute has shown evidence linking the antibody-catalyzed water-oxidation pathway of the human immune response to the production of ozone. In this system, ozone is produced by antibody-catalyzed production of trioxidane from water and neutrophil-produced singlet oxygen.

When inhaled, ozone reacts with compounds lining the lungs to form specific, cholesterol-derived metabolites that are thought to facilitate the build-up and pathogenesis of atherosclerotic plaques (a form of heart disease). These metabolites have been confirmed as naturally occurring in human atherosclerotic arteries and are categorized into a class of secosterols termed *atheronals*, generated by ozonolysis of cholesterol's double bond to form a 5,6 secosterol as well as a secondary condensation product via aldolization.

Ozone has been implicated to have an adverse effect on plant growth: "... ozone reduced total chlorophylls, carotenoid and carbohydrate concentration, and increased 1-aminocyclopropane-1-carboxylic acid (ACC) content and ethylene production. In treated plants, the ascorbate leaf pool was decreased, while lipid peroxidation and solute leakage were significantly higher than in ozone-free controls. The data indicated that ozone triggered protective mechanisms against oxidative stress in citrus."

Safety Regulations

Because of the strongly oxidizing properties of ozone, ozone is a primary irritant, affecting especially the eyes and respiratory systems and can be hazardous at even low concentrations. The Canadian Center for Occupation Safety and Health reports that:

"Even very low concentrations of ozone can be harmful to the upper respiratory tract and the lungs. The severity of injury depends on both by the concentration of ozone and the duration of exposure. Severe and permanent lung injury or death could result from even a very short-term exposure to relatively low concentrations."

To protect workers potentially exposed to ozone, U.S. Occupational Safety and Health Administration has established a permissible exposure limit (PEL) of 0.1 μmol/mol (29 CFR 1910.1000 table Z-1), calculated as an 8-hour time weighted average. Higher concentrations are especially hazardous and NIOSH has established an Immediately Dangerous to Life and Health Limit (IDLH) of 5 μmol/mol. Work environments where ozone is used or where it is likely to be produced should have adequate ventilation and it is prudent to have a monitor for ozone that will alarm if the concentration exceeds the OSHA PEL. Continuous monitors for ozone are available from several suppliers.

Elevated ozone exposure can occur on passenger aircraft, with levels depending on altitude and atmospheric turbulence. United States Federal Aviation Authority regulations set a limit of 250 nmol/mol with a maximum four-hour average of 100 nmol/mol. Some planes are equipped with ozone converters in the ventilation system to reduce passenger exposure.

Production

Ozone production demonstration, Fixed Nitrogen Research Laboratory, 1926

Ozone generators are used to produce ozone for cleaning air or remove smoke odors in unoccupied rooms. These ozone generators can produce over 3 g of ozone per hour. Ozone often forms in nature under conditions where O_2 will not react. Ozone used in industry is measured in μmol/mol (ppm, parts per million), nmol/mol (ppb, parts per billion), μg/m³, mg/h (milligrams per hour) or weight percent. The regime of applied concentrations ranges from 1 to 5% in air and from 6 to 14% in oxygen for older generation methods. New electrolytic methods can achieve up 20 to 30% dissolved ozone concentrations in output water.

Temperature and humidity play a large role in how much ozone is being produced using traditional generation methods such as corona discharge and ultraviolet light. Old generation methods will produce less than 50% its nominal capacity if operated with humid ambient air than when it operates in very dry air. New generators using electrolytic methods can achieve higher purity and dissolution through using water molecules as the source of ozone production.

Corona Discharge Method

This is the most common type of ozone generator for most industrial and personal uses. While variations of the "hot spark" coronal discharge method of ozone production exist, including medical grade and industrial grade ozone generators, these units usually work by means of a corona discharge tube. They are typically cost-effective and do not require an oxygen source other than the ambient air to produce ozone concentrations of 3–6%. Fluctuations in ambient air, due to weather or other environmental conditions, cause variability in ozone production. However, they also produce nitrogen oxides as a by-product. Use of an air dryer can reduce or eliminate nitric acid formation by removing water vapor and increase ozone production. Use of an oxygen concentrator can further increase the ozone production and further reduce the risk of nitric acid formation by removing not only the water vapor, but also the bulk of the nitrogen.

Ultraviolet Light

UV ozone generators, or vacuum-ultraviolet (VUV) ozone generators, employ a light source that generates a narrow-band ultraviolet light, a subset of that produced by the Sun. The Sun's UV sustains the ozone layer in the stratosphere of Earth.

While standard UV ozone generators tend to be less expensive, they usually produce ozone with a concentration of about 0.5% or lower. Another disadvantage of this method is that it requires the air (oxygen) to be exposed to the UV source for a longer amount of time, and any gas that is not exposed to the UV source will not be treated. This makes UV generators impractical for use in situations that deal with rapidly moving air or water streams (in-duct air sterilization, for example). Production of ozone is one of the potential dangers of ultraviolet germicidal irradiation. VUV ozone generators are used in swimming pool and spa applications ranging to millions of gallons of water. VUV ozone generators, unlike corona discharge generators, do not produce harmful nitrogen by-products and also unlike corona discharge systems, VUV ozone generators work extremely well in humid air environments. There is also not normally a need for expensive off-gas mechanisms, and no need for air driers or oxygen concentrators which require extra costs and maintenance.

Cold Plasma

In the cold plasma method, pure oxygen gas is exposed to a plasma created by dielectric barrier discharge. The diatomic oxygen is split into single atoms, which then recombine in triplets to form ozone.

Cold plasma machines utilize pure oxygen as the input source and produce a maximum concentration of about 5% ozone. They produce far greater quantities of ozone in a given space of time compared to ultraviolet production. However, because cold plasma ozone generators are very expensive, they are found less frequently than the previous two types.

The discharges manifest as filamentary transfer of electrons (micro discharges) in a gap between two electrodes. In order to evenly distribute the micro discharges, a dielectric insulator must be used to separate the metallic electrodes and to prevent arcing.

Some cold plasma units also have the capability of producing short-lived allotropes of oxygen which include O_4, O_5, O_6, O_7, etc. These species are even more reactive than ordinary O3.

Electrolytic

Electrolytic ozone generation (EOG) splits water molecules into H_2, O_2, and O_3. In most EOG methods, the hydrogen gas will be removed to leave oxygen and ozone as the only reaction products. Therefore, EOG can achieve higher dissolution in water without other competing gases found in corona discharge method, such as nitrogen gases present in ambient air. This method of generation can achieve concentrations of 20–30% and is independent of air quality because water is used as the source material. Production of ozone electrolytically is typically unfavorable because of the high overpotential required to produce ozone as compared to oxygen. This is why ozone is not produced during typical water electrolysis. However, it is possible to increase the overpotential of oxygen by careful catalyst selection such that ozone is preferentially produced under electrolysis. Catalysts typically chosen for this approach are lead dioxide or boron-doped diamond.

Special Considerations

Ozone cannot be stored and transported like other industrial gases (because it quickly decays into diatomic oxygen) and must therefore be produced on site. Available ozone generators vary in the arrangement and design of the high-voltage electrodes. At production capacities higher than 20 kg per hour, a gas/water tube heat-exchanger may be utilized as ground electrode and assembled with tubular high-voltage electrodes on the gas-side. The regime of typical gas pressures is around 2 bars (200 kPa) absolute in oxygen and 3 bars (300 kPa) absolute in air. Several megawatts of electrical power may be installed in large facilities, applied as one phase AC current at 50 to 8000 Hz and peak voltages between 3,000 and 20,000 volts. Applied voltage is usually inversely related to the applied frequency.

The dominating parameter influencing ozone generation efficiency is the gas temperature, which is controlled by cooling water temperature and/or gas velocity. The cooler the water, the better the ozone synthesis. The lower the gas velocity, the higher the concentration (but the lower the net ozone produced). At typical industrial conditions, almost 90% of the effective power is dissipated as heat and needs to be removed by a sufficient cooling water flow.

Because of the high reactivity of ozone, only a few materials may be used like stainless steel (quality 316L), titanium, aluminium (as long as no moisture is present), glass, polytetrafluorethylene, or polyvinylidene fluoride. Viton may be used with the restriction of constant mechanical forces and absence of humidity (humidity limitations apply depending on the formulation). Hypalon may be used with the restriction that no water come in contact with it, except for normal atmospheric levels. Embrittlement or shrinkage is the common mode of failure of elastomers with exposure to ozone. Ozone cracking is the common mode of failure of elastomer seals like O-rings.

Silicone rubbers are usually adequate for use as gaskets in ozone concentrations below 1 wt%, such as in equipment for accelerated aging of rubber samples.

Incidental Production

Ozone may be formed from O_2 by electrical discharges and by action of high energy electromagnetic radiation. Unsuppressed arcing in electrical contacts, motor bushes, or mechanical switches breaks down the chemical bonds of the atmospheric oxygen surrounding the contacts $[O_2 \rightarrow 2O]$. Free radicals of oxygen in and around the arc recombine to create ozone $[O_3]$. Certain electrical equipment generate significant levels of ozone. This is especially true of devices using high voltages, such as ionic air purifiers, laser printers, photocopiers, tasers and arc welders. Electric motors using brushes can generate ozone from repeated sparking inside the unit. Large motors that use brushes, such as those used by elevators or hydraulic pumps, will generate more ozone than smaller motors.

Ozone is similarly formed in the Catatumbo lightning storms phenomenon on the Catatumbo River in Venezuela, which helps to replenish ozone in the upper troposphere. It is the world's largest single natural generator of ozone, lending calls for it to be designated a UNESCO World Heritage Site.

Laboratory Production

In the laboratory, ozone can be produced by electrolysis using a 9 volt battery, a pencil graphite rod cathode, a platinum wire anode and a 3 molar sulfuric acid electrolyte. The half cell reactions taking place are:

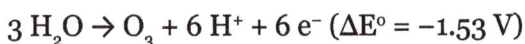

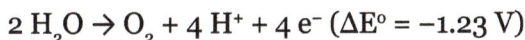

$$3 \ H_2O \rightarrow O_3 + 6 \ H^+ + 6 \ e^- \ (\Delta E^\circ = -1.53 \ V)$$

$$6 \ H^+ + 6 \ e^- \rightarrow 3 \ H_2 \ (\Delta E^\circ = 0 \ V)$$

$$2 \ H_2O \rightarrow O_2 + 4 \ H^+ + 4 \ e^- \ (\Delta E^\circ = -1.23 \ V)$$

In the net reaction, three equivalents of water are converted into one equivalent of ozone and three equivalents of hydrogen. Oxygen formation is a competing reaction.

It can also be generated by a high voltage arc. In its simplest form, high voltage AC, such as the output of a Neon-sign transformer is connected to two metal rods with the ends placed sufficiently close to each other to allow an arc. The resulting arc will convert atmospheric oxygen to ozone.

It is often desirable to contain the ozone. This can be done with an apparatus consisting of two concentric glass tubes sealed together at the top with gas ports at the top and bottom of the outer tube. The inner core should have a length of metal foil inserted into it connected to one side of the power source. The other side of the power source should be connected to another piece of foil wrapped around the outer tube. A source of dry O2 is applied to the bottom port. When high voltage is applied to the foil leads, electricity will discharge between the dry dioxygen in the middle and form O3 and O2 which will flow out the top port. The reaction can be summarized as follows:

$$3 \ O_2 - \textit{electricity} \rightarrow 2 \ O_3$$

Applications

Industry

The largest use of ozone is in the preparation of pharmaceuticals, synthetic lubricants, and many other commercially useful organic compounds, where it is used to sever carbon-carbon bonds. It can also be used for bleaching substances and for killing microorganisms in air and water sources. Many municipal drinking water systems kill bacteria with ozone instead of the more common chlorine. Ozone has a very high oxidation potential. Ozone does not form organochlorine compounds, nor does it remain in the water after treatment. Ozone can form the suspected carcinogen bromate in source water with high bromide concentrations. The U.S. Safe Drinking Water Act mandates that these systems introduce an amount of chlorine to maintain a minimum of 0.2 μmol/mol residual free chlorine in the pipes, based on results of regular testing. Where electrical power is abundant, ozone is a cost-effective method of treating water, since it is produced on demand and does not require transportation and storage of hazardous chemicals. Once it has decayed, it leaves no taste or odour in drinking water.

Although low levels of ozone have been advertised to be of some disinfectant use in residential homes, the concentration of ozone in dry air required to have a rapid, substantial effect on airborne pathogens exceeds safe levels recommended by the U.S. Occupational Safety and Health Administration and Environmental Protection Agency. Humidity control can vastly improve both the killing power of the ozone and the rate at which it decays back to oxygen (more humidity allows more effectiveness). Spore forms of most pathogens are very tolerant of atmospheric ozone in concentrations where asthma patients start to have issues.

Industrially, ozone is used to:

- Disinfect laundry in hospitals, food factories, care homes etc.;

- Disinfect water in place of chlorine

- Deodorize air and objects, such as after a fire. This process is extensively used in fabric restoration

- Kill bacteria on food or on contact surfaces;

- Sanitize swimming pools and spas

- Kill insects in stored grain

- Scrub yeast and mold spores from the air in food processing plants;

- Wash fresh fruits and vegetables to kill yeast, mold and bacteria;

- Chemically attack contaminants in water (iron, arsenic, hydrogen sulfide, nitrites, and complex organics lumped together as "colour");

- Provide an aid to flocculation (agglomeration of molecules, which aids in filtration, where the iron and arsenic are removed);

- Manufacture chemical compounds via chemical synthesis

- Clean and bleach fabrics (the former use is utilized in fabric restoration; the latter use is patented);

- Act as an antichlor in chlorine-based bleaching;

- Assist in processing plastics to allow adhesion of inks;

- Age rubber samples to determine the useful life of a batch of rubber;

- Eradicate water borne parasites such as *Giardia lamblia* and *Cryptosporidium* in surface water treatment plants.

Ozone is a reagent in many organic reactions in the laboratory and in industry. Ozonolysis is the cleavage of an alkene to carbonyl compounds.

Many hospitals around the world use large ozone generators to decontaminate operating rooms between surgeries. The rooms are cleaned and then sealed airtight before being filled with ozone which effectively kills or neutralizes all remaining bacteria.

Ozone is used as an alternative to chlorine or chlorine dioxide in the bleaching of wood pulp. It is often used in conjunction with oxygen and hydrogen peroxide to eliminate the need for chlorine-containing compounds in the manufacture of high-quality, white paper.

Ozone can be used to detoxify cyanide wastes (for example from gold and silver mining) by oxidising cyanide to cyanate and eventually to carbon dioxide.

Consumers

Devices generating high levels of ozone, some of which use ionization, are used to sanitize and deodorize uninhabited buildings, rooms, ductwork, woodsheds, boats and other vehicles.

In the U.S., air purifiers emitting low levels of ozone have been sold. This kind of air purifier is sometimes claimed to imitate nature's way of purifying the air without filters and to sanitize both it and household surfaces. The United States Environmental Protection Agency (EPA) has declared that there is "evidence to show that at concentrations that do not exceed public health standards, ozone is not effective at removing many odor-causing chemicals" or "viruses, bacteria, mold, or other biological pollutants". Furthermore, its report states that "results of some controlled studies show that concentrations of ozone considerably higher than these [human safety] standards are possible even when a user follows the manufacturer's operating instructions". A couple kept repeating health claims for the generator they sold, without supporting scientific studies. In 1998, a federal jury convicted them, among others things, of illegally distributing an ozone generator and of wire fraud.

Ozonated water is used to launder clothes and to sanitize food, drinking water, and surfaces in the home. According to the U.S. Food and Drug Administration (FDA), it is "amending the food additive regulations to provide for the safe use of ozone in gaseous and aqueous phases as an antimicrobial agent on food, including meat and poultry." Studies at California Polytechnic University demonstrated that 0.3 μmol/mol levels of ozone dissolved in filtered tapwater can produce a reduction of more than 99.99% in such food-borne microorganisms as salmonella, *E. coli* 0157:H7 and *Campylobacter*. This quantity is 20,000 times the WHO-recommended limits stated above. Ozone can be used to remove pesticide residues from fruits and vegetables.

Ozone is used in homes and hot tubs to kill bacteria in the water and to reduce the amount of chlorine or bromine required by reactivating them to their free state. Since ozone does not remain in the water long enough, ozone by itself is ineffective at preventing cross-contamination among bathers and must be used in conjunction with halogens. Gaseous ozone created by ultraviolet light or by corona discharge is injected into the water.

Ozone is also widely used in treatment of water in aquariums and fish ponds. Its use can minimize bacterial growth, control parasites, eliminate transmission of some diseases, and reduce or eliminate "yellowing" of the water. Ozone must not come in contact with fish's gill structures. Natural salt water (with life forms) provides enough "instantaneous demand" that controlled amounts of ozone activate bromide ion to hypobromous acid, and the ozone entirely decays in a few seconds to minutes. If oxygen fed ozone is used, the water will be higher in dissolved oxygen, fish's gill structures will atrophy and they will become dependent on higher dissolved oxygen levels.

Aquaculture

Ozonation - a process of infusing water with ozone - can be used in aquaculture to facilitate organic breakdown. Ozone is also added to recirculating systems to reduce nitrite levels through conversion into nitrate. If nitrite levels in the water are high, nitrites will also accumulate in the blood and tissues of fish, where it interferes with oxygen transport (it causes oxidation of the heme-group of haemoglobin from ferrous (Fe^{2+}) to ferric (Fe^{3+}), making haemoglobin unable to bind O2). Despite these apparent positive effects, ozone use in recirculation systems has been linked to reducing the level of bioavailable iodine in salt water systems, resulting in iodine deficiency symptoms such as goitre and decreased growth in Senegalese sole (Solea senegalensis) larvae.

Ozonate seawater is used for surface disinfection of haddock and Atlantic halibut eggs against nodavirus. Nodavirus is a lethal and vertically transmitted virus which causes severe mortality in fish. Haddock eggs should not be treated with high ozone level as eggs so treated did not hatch and died after 3–4 days.

Agriculture

Ozone application on freshly cut pineapple and banana shows increase in flavonoids and total phenol contents when exposure is up to 20 minutes. Decrease in ascorbic acid (one form of vitamin C) content is observed but the positive effect on total phenol content and flavonoids can overcome the negative effect. Tomatoes upon treatment with ozone shows an increase in β-carotene, lutein and lycopene. However, ozone application on strawberries in pre-harvest period shows decrease in ascorbic acid content.

Ozone facilitates the extraction of some heavy metals from soil using EDTA. EDTA forms strong, water-soluble coordination compounds with some heavy metals (Pb, Zn) thereby making it possible to dissolve them out from contaminated soil. If contaminated soil is pre-treated with ozone, the extraction efficacy of Pb, Am and Pu increases by 11.0–28.9%, 43.5% and 50.7% respectively.

Medical

Various therapeutic uses for Ozone have been proposed, but are not supported by peer-reviewed evidence and generally considered alternative medicine.

Nitrous Oxide

Nitrous oxide, commonly known as laughing gas, nitrous, nitro, or NOS is a chemical compound with the formula N2O. It is an oxide of nitrogen. At room temperature, it is a colorless, non-flammable gas, with a slightly sweet odor and taste. It is used in surgery and dentistry for its anaesthetic and analgesic effects. It is known as "laughing gas" due to the euphoric effects of inhaling it, a property that has led to its recreational use as a dissociative anaesthetic. It is also used as an oxidizer in rocket propellants, and in motor racing to increase the power output of engines. At elevated temperatures, nitrous oxide is a powerful oxidizer similar to molecular oxygen.

Nitrous oxide gives rise to nitric oxide (NO) on reaction with oxygen atoms, and this NO in turn reacts with ozone. As a result, it is the main naturally occurring regulator of stratospheric ozone. It is also a major greenhouse gas and air pollutant. Considered over a 100-year period, it is calculated to have between 265 and 310 times more impact per unit mass (global-warming potential) than carbon dioxide.

It is on the WHO Model List of Essential Medicines, the most important medications needed in a health system.

Applications

Rocket Motors

Nitrous oxide can be used as an oxidizer in a rocket motor. This has the advantages over other oxidisers in that it is not only non-toxic, but also, due to its stability at room temperature, easy to store and relatively safe to carry on a flight. As a secondary benefit it can be readily decomposed to form breathing air. Its high density and low storage pressure (when maintained at low temperature) enable it to be highly competitive with stored high-pressure gas systems.

In a 1914 patent, American rocket pioneer Robert Goddard suggested nitrous oxide and gasoline as possible propellants for a liquid-fuelled rocket. Nitrous oxide has been the oxidiser of choice in several hybrid rocket designs (using solid fuel with a liquid or gaseous oxidizer). The combination of nitrous oxide with hydroxyl-terminated polybutadiene fuel has been used by SpaceShipOne and others. It is also notably used in amateur and high power rocketry with various plastics as the fuel.

Nitrous oxide can also be used in a monopropellant rocket. In the presence of a heated catalyst, N2O will decompose exothermically into nitrogen and oxygen, at a temperature of approximately 1,070 °F (577 °C). Because of the large heat release, the catalytic action rapidly becomes secondary as thermal autodecomposition becomes dominant. In a vacuum thruster, this can provide a monopropellant specific impulse (I_{sp}) of as much as 180 s. While noticeably less than the I_{sp} available from hydrazine thrusters (monopropellant or bipropellant with dinitrogen tetroxide), the decreased toxicity makes nitrous oxide an option worth investigating.

Nitrous oxide is said to deflagrate somewhere around 600 °C (1,112 °F) at a pressure of 21 atmospheres. At 600 psi for example, the required ignition energy is only 6 joules, whereas N2O at 130 psi a 2500-joule ignition energy input is insufficient.

Internal Combustion Engine

In vehicle racing, nitrous oxide (often referred to as just "nitrous") allows the engine to burn more fuel by providing more oxygen than air alone, resulting in a more powerful combustion.The gas itself is not flammable at a low pressure/temperature, but it delivers more oxygen than atmospheric air by breaking down at elevated temperatures. Therefore, it is often mixed with another fuel that is easier to deflagrate. Nitrous oxide is a strong oxidant roughly equivalent to hydrogen peroxide and much stronger than oxygen gas.

Nitrous oxide is stored as a compressed liquid; the evaporation and expansion of liquid nitrous oxide in the intake manifold causes a large drop in intake charge temperature, resulting in a denser charge, further allowing more air/fuel mixture to enter the cylinder. Nitrous oxide is sometimes injected into (or prior to) the intake manifold, whereas other systems directly inject right before the cylinder (direct port injection) to increase power.

The technique was used during World War II by Luftwaffe aircraft with the GM-1 system to boost the power output of aircraft engines. Originally meant to provide the Luftwaffe standard aircraft with superior high-altitude performance, technological considerations limited its use to extremely high altitudes. Accordingly, it was only used by specialized planes like high-altitude reconnaissance aircraft, high-speed bombers, and high-altitude interceptor aircraft. It could sometimes be found on Luftwaffe aircraft also fitted with another engine-boost system, MW 50, a form of water injection for aviation engines that used methanol for its boost capabilities.

One of the major problems of using nitrous oxide in a reciprocating engine is that it can produce enough power to damage or destroy the engine. Very large power increases are possible, and if the mechanical structure of the engine is not properly reinforced, the engine may be severely damaged or destroyed during this kind of operation. It is very important with nitrous oxide augmentation of petrol engines to maintain proper operating temperatures and fuel levels to prevent "pre-ignition", or "detonation" (sometimes referred to as "knock"). Most problems that are associated with nitrous do not come from mechanical failure due to the power increases. Since nitrous allows a much denser charge into the cylinder it dramatically increases cylinder pressures. The increased pressure and temperature can cause problems such as melting the piston or valves. It may also crack or warp the piston or head and cause pre-ignition due to uneven heating.

Automotive-grade liquid nitrous oxide differs slightly from medical-grade nitrous oxide. A small amount of sulfur dioxide (SO_2) is added to prevent substance abuse. Multiple washes through a base (such as sodium hydroxide) can remove this, decreasing the corrosive properties observed when SO_2 is further oxidised during combustion into sulfuric acid, making emissions cleaner.

Aerosol Propellant

The gas is approved for use as a food additive (also known as E942), specifically as an aerosol spray propellant. Its most common uses in this context are in aerosol whipped cream canisters, cooking sprays, and as an inert gas used to displace oxygen, to inhibit bacterial growth, when filling packages of potato chips and other similar snack foods.

The gas is extremely soluble in fatty compounds. In aerosol whipped cream, it is dissolved in the fatty cream until it leaves the can, when it becomes gaseous and thus creates foam. Used in this way, it produces whipped cream four times the volume of the liquid, whereas whipping air into cream only produces twice the volume. If air were used as a propellant, oxygen would accelerate rancidification of the butterfat; nitrous oxide inhibits such degradation. Carbon dioxide cannot be used for whipped cream because it is acidic in water, which would curdle the cream and give it a seltzer-like "sparkling" sensation.

However, the whipped cream produced with nitrous oxide is unstable and will return to a more liquid state within half an hour to one hour. Thus, the method is not suitable for decorating food that will not be immediately served.

Similarly, cooking spray, which is made from various types of oils combined with lecithin (an emulsifier), may use nitrous oxide as a propellant; other propellants used in cooking spray include food-grade alcohol and propane.

Users of nitrous oxide often obtain it from whipped cream dispensers that use nitrous oxide as a propellant, for recreational use as a euphoria-inducing inhalant drug. It is not harmful in small doses, but risks due to lack of oxygen do exist.

Medicine

Medical grade N2O tanks used in dentistry.

Nitrous oxide has been used for anaesthesia in dentistry since December 1844, where Horace

Wells made the first 12–15 dental operations with the gas in Hartford. Its debut as a generally accepted method, however, came in 1863, when Gardner Quincy Colton introduced it more broadly at all the Colton Dental Association clinics, that he founded in New Haven and New York City. The first devices used in dentistry to administer the gas, known as Nitrous Oxide inhalers, were designed in a very simple way with the gas stored and breathed through a breathing bag made of rubber cloth, without a scavenger system and flowmeter, and with no addition of oxygen/air. Today these simple and somewhat unreliable inhalers have been replaced by the more modern relative analgesia machine, which is an automated machine designed to deliver a precisely dosed and breath-actuated flow of nitrous oxide mixed with oxygen, for the patient to inhale safely. The machine used in dentistry is designed as a simplified version of the larger anaesthetic machine used by hospitals, as it doesn't feature the additional anaesthetic vaporiser and medical ventilator. The purpose of the machine allows for a simpler design, as it only delivers a mixture of nitrous oxide and oxygen for the patient to inhale, in order to depress the feeling of pain while keeping the patient in a conscious state.

Relative analgesia machines typically feature a constant-supply flowmeter, which allow the proportion of nitrous oxide and the combined gas flow rate to be individually adjusted. The gas is administered by dentists through a demand-valve inhaler over the nose, which will only release gas when the patient inhales through the nose. Because nitrous oxide is minimally metabolised in humans (with a rate of 0.004%), it retains its potency when exhaled into the room by the patient, and can pose an intoxicating and prolonged exposure hazard to the clinic staff if the room is poorly ventilated. Where nitrous oxide is administered, a continuous-flow fresh-air ventilation system or nitrous scavenger system is used to prevent a waste-gas buildup.

Hospitals administer nitrous oxide as one of the anaesthetic drugs delivered by anaesthetic machines. Nitrous oxide is a weak general anaesthetic, and so is generally not used alone in general anaesthesia. In general anaesthesia it is used as a carrier gas in a 2:1 ratio with oxygen for more powerful general anaesthetic drugs such as sevoflurane or desflurane. It has a minimum alveolar concentration of 105% and a blood/gas partition coefficient of 0.46.

The medical grade gas tanks, with the tradename Entonox and Nitronox contain a mixture with 50%, but this will normally be diluted to a lower percentage upon the operational delivery to the patient. Inhalation of nitrous oxide is frequently used to relieve pain associated with childbirth, trauma, oral surgery, and acute coronary syndrome (includes heart attacks). Its use during labour has been shown to be a safe and effective aid for birthing women. Its use for acute coronary syndrome is of unknown benefit.

In Britain and Canada, Entonox and Nitronox are commonly used by ambulance crews (including unregistered practitioners) as a rapid and highly effective analgesic gas.

Nitrous oxide has been shown to be effective in treating a number of addictions, including alcohol withdrawal.

Nitrous oxide is also gaining interest as a substitute gas for carbon dioxide in laparoscopic surgery. It has been found to be as safe as carbon dioxide with better pain relief.

Recreational use

Food grade N2O whippets (above) and cracker (below)—can be used for recreational purposes

Nitrous oxide can cause analgesia, depersonalisation, derealisation, dizziness, euphoria, and some sound distortion. Research has also found that it increases suggestibility and imagination. Inhalation of nitrous oxide for recreational use, with the purpose of causing euphoria and/or slight hallucinations, began as a phenomenon for the British upper class in 1799, known as "laughing gas parties". Until at least 1863, a low availability of equipment to produce the gas, combined with a low usage of the gas for medical purposes, meant it was a relatively rare phenomenon that mainly happened among students at medical universities. When equipment became more widely available for dentistry and hospitals, most countries also restricted the legal access to buy pure nitrous oxide gas cylinders to those sectors. Despite only medical staff and dentists today being legally allowed to buy the pure gas, a Consumers Union report from 1972 found that the use of the gas for recreational purpose was [then] still taking place, based upon reports of its use in Maryland 1971, Vancouver 1972, and a survey made by Dr. Edward J. Lynn of its non-medical use in Michigan 1970.

It was not uncommon [in the interviews] to hear from individuals who had been to parties where a professional (doctor, nurse, scientist, inhalation therapist, researcher) had provided nitrous oxide. There also were those who work in restaurants who used the N2O stored in tanks for the preparation of whip cream. Reports were received from people who used the gas contained in aerosol cans both of food and non-food products. At a recent rock festival nitrous oxide was widely sold for 25 cents a balloon. Contact was made with a "mystical-religious" group that used the gas to accelerate arriving at their transcendental-meditative state of choice. Although a few, more sophisticated users employed nitrous oxide-oxygen mixes with elaborate equipment, most users used balloons or plastic bags. They either held a breath of N_2O or rebreathed the gas. There were no adverse effects reported in the more than one hundred individuals surveyed.

In Australia, nitrous oxide bulbs are known as nangs, possibly derived from the sound distortion perceived by consumers.

In the United Kingdom, nitrous oxide is used by almost half a million young people at nightspots, festivals and parties. In August 2015, the London Borough of Lambeth Council banned the use of the drug for recreational purposes, making offenders liable to an on-the-spot fine of up to £1,000.

Mechanism of Action

The pharmacological mechanism of action of N2O in medicine is not fully known. However, it has been shown to directly modulate a broad range of ligand-gated ion channels, and this likely plays a major role in many of its effects. It moderately blocks NMDA and β_2-subunit-containing nACh channels, weakly inhibits AMPA, kainate, $GABA_C$, and $5\text{-}HT_3$ receptors, and slightly potentiates $GABA_A$ and glycine receptors. It has also been shown to activate two-pore-domain K+channels. While N2O affects quite a few ion channels, its anaesthetic, hallucinogenic, and euphoriant effects are likely caused predominantly or fully via inhibition of NMDA receptor-mediated currents. In addition to its effects on ion channels, N2O may act to imitate nitric oxide (NO) in the central nervous system, and this may be related to its analgesic and anxiolytic properties.

Anxiolytic Effect

In behavioural tests of anxiety, a low dose of N2O is an effective anxiolytic, and this anti-anxiety effect is associated with enhanced activity of $GABA_A$ receptors, as it is partially reversed by benzodiazepine receptor antagonists. Mirroring this, animals which have developed tolerance to the anxiolytic effects of benzodiazepines are partially tolerant to N2O. Indeed, in humans given 30% N2O, benzodiazepine receptor antagonists reduced the subjective reports of feeling "high", but did not alter psychomotor performance, in human clinical studies.

Analgesic Effect

The analgesic effects of N2O are linked to the interaction between the endogenous opioid system and the descending noradrenergic system. When animals are given morphine chronically they develop tolerance to its pain-killing effects, and this also renders the animals tolerant to the analgesic effects of N2O. Administration of antibodies which bind and block the activity of some endogenous opioids (not β-endorphin) also block the antinociceptive effects of N2O. Drugs which inhibit the breakdown of endogenous opioids also potentiate the antinociceptive effects of N2O. Several experiments have shown that opioid receptor antagonists applied directly to the brain block the antinociceptive effects of N2O, but these drugs have no effect when injected into the spinal cord.

Conversely, α_2-adrenoceptor antagonists block the pain reducing effects of N2O when given directly to the spinal cord, but not when applied directly to the brain. Indeed, α_{2B}-adrenoceptor knockout mice or animals depleted in norepinephrine are nearly completely resistant to the antinociceptive effects of N2O. Apparently N2O-induced release of endogenous opioids causes disinhibition of brain stem noradrenergic neurons, which release norepinephrine into the spinal cord and inhibit pain signalling. Exactly how N2O causes the release of endogenous opioid peptides is still uncertain.

Euphoric Effect

In rats, N2O stimulates the mesolimbic reward pathway via inducing dopamine release and activating dopaminergic neurons in the ventral tegmental area and nucleus accumbens, presumably through antagonisation of NMDA receptors localised in the system. This action has been implicated in its euphoric effects, and notably, appears to augment its analgesic properties as well.

However, it is remarkable that in mice, N2O blocks amphetamine-induced carrier-mediated dopamine release in the nucleus accumbens and behavioural sensitisation, abolishes the conditioned place preference (CPP) of cocaine and morphine, and does not produce reinforcing (or aversive) effects of its own. Studies on CPP of N2O in rats is mixed, consisting of reinforcement, aversion, and no change. In contrast, it is a positive reinforcer in squirrel monkeys, and is well known as a drug of abuse in humans. These discrepancies in response to N2O may reflect species variation or methodological differences. In human clinical studies, N2O was found to produce mixed responses similarly to rats, reflecting high subjective individual variability.

Neurotoxicity and Neuroprotection

Like other NMDA antagonists, N2O was suggested to produce neurotoxicity in the form of Olney's lesions in rodents upon prolonged (several hour) exposure. However, new research has arisen suggesting that Olney's lesions do not occur in humans, and similar drugs like ketamine are now believed not to be acutely neurotoxic. It has been argued that, because N2O has a very short duration under normal circumstances, it is less likely to be neurotoxic than other NMDA antagonists. Indeed, in rodents, short-term exposure results in only mild injury that is rapidly reversible, and permanent neuronal death only occurs after constant and sustained exposure. Nitrous oxide may also cause neurotoxicity after extended exposure because of hypoxia. This is especially true of non-medical formulations such as whipped-cream chargers (also known as "whippets" or "nangs"), which are not necessarily mixed with oxygen.

Additionally, nitrous oxide depletes vitamin B12 levels. This can cause serious neurotoxicity with even acute use if the user has preexisting vitamin B12 deficiency.

Nitrous oxide at 75-vol% reduce ischemia-induced neuronal death induced by occlusion of the middle cerebral artery in rodents, and decrease NMDA-induced Ca2+ influx in neuronal cell cultures, a critical event involved in excitotoxicity.

Safety

The major safety hazards of nitrous oxide come from the fact that it is a compressed liquefied gas, an asphyxiation risk, and a dissociative anaesthetic. Exposure to nitrous oxide causes short-term decreases in mental performance, audiovisual ability, and manual dexterity. Abusing nitrous oxide can lead to oxygen deprivation resulting in loss of blood pressure, fainting and even heart attacks.

Long-term exposure can cause vitamin B_{12} deficiency, numbness, reproductive side effects (in pregnant females), and other problems. The National Institute for Occupational Safety and Health recommends that workers' exposure to nitrous oxide should be controlled during the administration of anaesthetic gas in medical, dental, and veterinary operators. People can be exposed to nitrous oxide in the workplace by breathing it in or getting the liquid on their skin or in their eyes. The National Institute for Occupational Safety and Health (NIOSH) has set a Recommended exposure limit (REL) of 25 ppm (46 mg/m^3) exposure to waste anaesthetic.

Chemical/Physical

At room temperature (20 °C (68 °F)) the saturated vapour pressure is 50.525 bar, rising up to 72.45 bar at 36.4 °C (97.5 °F)—the critical temperature. The pressure curve is thus unusually sensitive to temperature. Liquid nitrous oxide acts as a good solvent for many organic compounds; liquid mixtures may form shock sensitive explosives.

As with many strong oxidisers, contamination of parts with fuels have been implicated in rocketry accidents, where small quantities of nitrous/fuel mixtures explode due to "water hammer"-like effects (sometimes called "dieseling"—heating due to adiabatic compression of gases can reach decomposition temperatures). Some common building materials such as stainless steel and aluminium can act as fuels with strong oxidisers such as nitrous oxide, as can contaminants, which can ignite due to adiabatic compression.

There have also been accidents where nitrous oxide decomposition in plumbing has led to the explosion of large tanks.

Biological

Nitrous oxide inactivates the cobalamin form of vitamin B_{12} by oxidation. Symptoms of vitamin B_{12} deficiency, including sensory neuropathy, myelopathy, and encephalopathy, can occur within days or weeks of exposure to nitrous oxide anaesthesia in people with subclinical vitamin B_{12} deficiency. Symptoms are treated with high doses of vitamin B_{12}, but recovery can be slow and incomplete. People with normal vitamin B_{12} levels have stores to make the effects of nitrous oxide insignificant, unless exposure is repeated and prolonged (nitrous oxide abuse). Vitamin B_{12} levels should be checked in people with risk factors for vitamin B_{12} deficiency prior to using nitrous oxide anaesthesia.

A study of workers and several experimental animal studies indicate that adverse reproductive effects for pregnant females may also result from chronic exposure to nitrous oxide.

Nitrous oxide reductase is an important enzyme which limits the emission of the gas to the atmosphere.

Environmental

N2O is a greenhouse gas with a large global warming potential (GWP). When compared to carbon dioxide (CO2), N2O has 298 times the ability per molecule of gas to trap heat in the atmosphere. N2O is produced naturally in the soil during the microbial processes of nitrification and denitrification.

The United States of America signed and ratified the United Nations Framework Convention on Climate Change (UNFCCC) in 1992, agreeing to inventory and assess the various sources of greenhouse gases that contribute to climate change. The agreement requires parties to "develop, periodically update, publish and make available... national inventories of anthropogenic emissions by sources and removals by sinks of all greenhouse gases not controlled by the Montreal Protocol, using comparable methodologies...". In response to this agreement, the U.S. is obligated to inventory anthropogenic emissions by sources and sinks, of which agriculture is a key contributor. In 2008, agriculture contributed 6.1% of the total U.S. greenhouse gas emissions and cropland contributed nearly 69% of total direct nitrous oxide (N2O) emissions. Additionally, estimated emissions from agricultural soils were 6% higher in 2008 than 1990.

According to 2006 data from the United States Environmental Protection Agency, industrial sources make up only about 20% of all anthropogenic sources, and include the production of nylon, and the burning of fossil fuel in internal combustion engines. Human activity is thought to account for 30%; tropical soils and oceanic release account for 70%. However, a 2008 study by Nobel Laureate Paul Crutzen suggests that the amount of nitrous oxide release attributable to agricultural nitrate fertilizers has been seriously underestimated, most of which would presumably come under soil and oceanic release in the Environmental Protection Agency data. Atmospheric levels have risen by more than 15% since 1750. Nitrous oxide also causes ozone depletion. A new study suggests that N_2O emission currently is the single most important ozone-depleting substance (ODS) emission and is expected to remain the largest throughout the 21st century.

Production

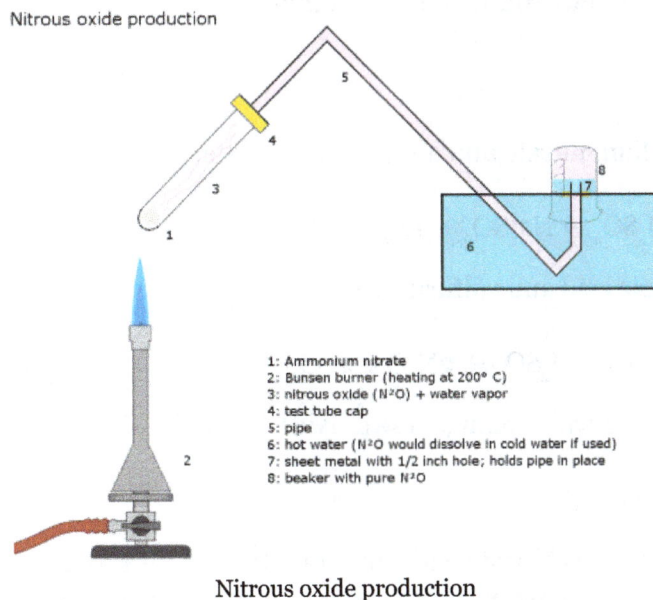

Nitrous oxide production

1: Ammonium nitrate
2: Bunsen burner (heating at 200° C)
3: nitrous oxide (N_2O) + water vapor
4: test tube cap
5: pipe
6: hot water (N_2O would dissolve in cold water if used)
7: sheet metal with 1/2 inch hole; holds pipe in place
8: beaker with pure N_2O

Nitrous oxide production

Nitrous oxide is most commonly prepared by careful heating of ammonium nitrate, which decomposes into nitrous oxide and water vapour. The addition of various phosphates favours formation of a purer gas at slightly lower temperatures. One of the earliest commercial producers was George Poe in Trenton, New Jersey.

$$NH_4NO_3 (s) \rightarrow 2H_2O (g) + N_2O (g)$$

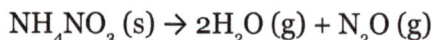

This reaction occurs between 170 and 240 °C (338 and 464 °F), temperatures where ammonium nitrate is a moderately sensitive explosive and a very powerful oxidizer. Above 240 °C (464 °F) the exothermic reaction may accelerate to the point of detonation, so the mixture must be cooled to avoid such a disaster. Superheated steam is used to reach reaction temperature in some turnkey production plants.

Downstream, the hot, corrosive mixture of gases must be cooled to condense the steam, and filtered to remove higher oxides of nitrogen. Ammonium nitrate smoke, as an extremely persistent colloid, will also have to be removed. The cleanup is often done in a train of three gas washes; namely base, acid and base again. Any significant amounts of nitric oxide (NO) may not necessarily be absorbed directly by the base (sodium hydroxide) washes.

The nitric oxide impurity is sometimes chelated out with ferrous sulfate, reduced with iron metal, or oxidised and absorbed in base as a higher oxide. The first base wash may (or may not) react out much of the ammonium nitrate smoke. However, this reaction generates ammonia gas, which may have to be absorbed in the acid wash.

As a Byproduct

The synthesis of adipic acid; one of the two reactants used in nylon manufacture, produces nitrogen oxides including nitric oxides. This might become a major commercial source, but will require the removal of higher oxides of nitrogen and organic impurities. Currently much of the gas is decomposed before release for environmental protection.

Other Routes

Heating a mixture of sodium nitrate and ammonium sulfate.

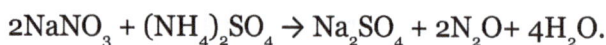

$$2NaNO_3 + (NH_4)_2SO_4 \rightarrow Na_2SO_4 + 2N_2O + 4H_2O.$$

The reaction of urea, nitric acid and sulfuric acid

$$2(NH_2)_2CO + 2HNO_3 + H_2SO_4 \rightarrow 2N_2O + 2CO_2 + (NH_4)_2SO_4 + 2H_2O.$$

Direct oxidation of ammonia with a manganese dioxide-bismuth oxide catalyst: cf. Ostwald process.

$$2NH_3 + 2O_2 \rightarrow N_2O + 3H_2O$$

Reacting Hydroxylammonium chloride with sodium nitrite. If the nitrite is added to the hydroxylamine solution, the only remaining by-product is salt water. However, if the hydroxylamine solution is added to the nitrite solution (nitrite is in excess), then toxic higher oxides of nitrogen are also formed.

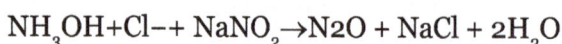

$$NH_3OH+Cl- + NaNO_2 \rightarrow N2O + NaCl + 2H_2O$$

Reacting HNO_3 with SnCl2 and HCl:

$$2\ HNO_3 + 8HCl + 4SnCl_2 \rightarrow 5H_2O + 4SnCl_4 + N_2O$$

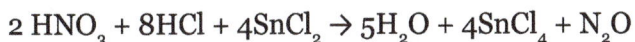

Hyponitrous acid decomposes to N_2O and water with a half-life of 16 days at 25 °C at pH 1–3.

$$H_2N_2O_2 \rightarrow H_2O + N_2O$$

Soil

Of the entire anthropogenic N_2O emission (5.7 teragrams N_2O-N per year), agricultural soils provide 3.5 teragrams N_2O–N per year. Nitrous oxide is produced naturally in the soil during the microbial processes of nitrification, denitrification, nitrifier denitrification and others:

- aerobic autotrophic nitrification, the stepwise oxidation of ammonia (NH_3) to nitrite (NO_2^-) and to nitrate (NO_3^-) (e.g., Kowalchuk and Stephen, 2001),

- anaerobic heterotrophic denitrification, the stepwise reduction of NO_3^- to NO_2^-, nitric oxide (NO), N_2O and ultimately N_2, where facultative anaerobe bacteria use NO_3^- as an electron acceptor in the respiration of organic material in the condition of insufficient oxygen (O_2) (e.g. Knowles, 1982), and

- nitrifier denitrification, which is carried out by autotrophic NH_3–oxidizing bacteria and the pathway whereby ammonia (NH_3) is oxidised to nitrite (NO_2^-), followed by the reduction of NO_2^- to nitric oxide (NO), N_2O and molecular nitrogen (N_2) (e.g., Webster and Hopkins, 1996; Wrage et al., 2001).

- Other N2O production mechanisms include heterotrophic nitrification (Robertson and Kuenen, 1990), aerobic denitrification by the same heterotrophic nitrifiers (Robertson and Kuenen, 1990), fungal denitrification (Laughlin and Stevens, 2002), and non-biological process chemodenitrification (e.g. Chalk and Smith, 1983; Van Cleemput and Baert, 1984; Martikainen and De Boer, 1993; Daum and Schenk, 1998; Mørkved et al., 2007).

Soil N2O emissions are reported to be controlled by soil chemical and physical properties such as the availability of mineral N, soil pH, organic matter availability, and soil type, and climate related soil properties such as soil temperature and soil water content (e.g., Mosier, 1994; Bouwman, 1996; Beauchamp, 1997; Yamulki et al. 1997; Dobbie and Smith, 2003; Smith et al. 2003; Dalal et al. 2003).

Properties and Reactions

Nitrous oxide is a colourless, non-toxic gas with a faint, sweet odour.

Nitrous oxide supports combustion by releasing the dipolar bonded oxygen radical,thus it can relight a glowing splint.

N_2O is inert at room temperature and has few reactions. At elevated temperatures, its reactivity increases. For example, nitrous oxide reacts with $NaNH_2$ at 460 K (187 °C) to give NaN_3:

$$2NaNH_2 + N_2O \rightarrow NaN_3 + NaOH + NH_3$$

The above reaction is the route adopted by the commercial chemical industry to produce azide salts, which are used as detonators.

Occurrence

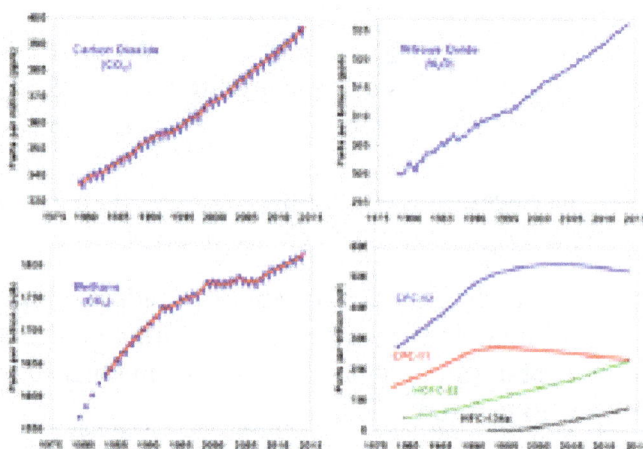

Greenhouse gas trends.

Nitrous oxide is emitted by bacteria in soils and oceans, and is thus a part of Earth's atmosphere. Agriculture is the main source of human-produced nitrous oxide: cultivating soil, the use of nitrogen fertilisers, and animal waste handling can all stimulate naturally occurring bacteria to produce more nitrous oxide. The livestock sector (primarily cows, chickens, and pigs) produces 65% of human-related nitrous oxide. Industrial sources make up only about 20% of all anthropogenic sources, and include the production of nylon, and the burning of fossil fuel in internal combustion engines. Human activity is thought to account for 40%; tropical soils and oceanic release account for the rest.

Nitrous oxide reacts with ozone in the stratosphere. Nitrous oxide is the main naturally occurring regulator of stratospheric ozone. Nitrous oxide is a major greenhouse gas. Considered over a 100-year period, it has 298 times more impact per unit weight than carbon dioxide. Thus, despite its low concentration, nitrous oxide is the fourth largest contributor to these greenhouse gases. It ranks behind water vapour, carbon dioxide, and methane. Control of nitrous oxide is part of efforts to curb greenhouse gas emissions.

History

The gas was first synthesised by English natural philosopher and chemist Joseph Priestley in 1772, who called it *phlogisticated nitrous air*. Priestley published his discovery in the book *Experiments and Observations on Different Kinds of Air (1775)*, where he described how to produce the preparation of "nitrous air diminished", by heating iron filings dampened with nitric acid.

Early Use

The first important use of nitrous oxide was made possible by Thomas Beddoes and James Watt, who worked together to publish the book *Considerations on the Medical Use and on the Production of Factitious Airs (1794)*. This book was important for two reasons. First, James Watt had invented a novel machine to produce "Factitious Airs" (i.e. nitrous oxide) and a novel "breathing apparatus" to inhale the gas. Second, the book also presented the new medical theories by Thomas Beddoes, that tuberculosis and other lung diseases could be treated by inhalation of "Factitious Airs".

The machine to produce "Factitious Airs" had three parts: A furnace to burn the needed material, a vessel with water where the produced gas passed through in a spiral pipe (for impurities to be "washed off"), and finally the gas cylinder with a gasometer where the gas produced, "air", could be tapped into portable air bags (made of airtight oily silk). The breathing apparatus consisted of one of the portable air bags connected with a tube to a mouthpiece. With this new equipment being engineered and produced by 1794, the way was paved for clinical trials, which began when Thomas Beddoes in 1798 established the *"Pneumatic Institution for Relieving Diseases by Medical Airs"* in Hotwells (Bristol). In the basement of the building, a large-scale machine was producing the gases under the supervision of a young Humphry Davy, who was encouraged to experiment with new gases for patients to inhale. The first important work of Davy was examination of the nitrous oxide, and the publication of his results in the book: *Researches, Chemical and Philosophical (1800)*.

Despite Davy's discovery that inhalation of nitrous oxide could relieve a conscious person from pain, another 44 years elapsed before doctors attempted to use it for anaesthesia. The use of nitrous oxide as a recreational drug at "laughing gas parties", primarily arranged for the British upper class, became an immediate success beginning in 1799. While the effects of the gas generally make the user appear stuporous, dreamy and sedated, some people also "get the giggles" in a state of euphoria, and frequently erupt in laughter.

Anaesthetic Use

The first time nitrous oxide was used as an anaesthetic drug in the treatment of a patient was when dentist Horace Wells, with assistance by Gardner Quincy Colton and John Mankey Riggs, demonstrated insensitivity to pain from a dental extraction on 11 December 1844. In the following weeks, Wells treated the first 12–15 patients with nitrous oxide in Hartford, and according to his own record only failed in two cases. In spite of these convincing results being reported by Wells to the medical society in Boston already in December 1844, this new method was not immediately adopted by other dentists. The reason for this was most likely that Wells, in January 1845 at his first public demonstration to the medical faculty in Boston, had been partly unsuccessful, leaving his colleagues doubtful regarding its efficacy and safety. The method did not come into general use until 1863, when Gardner Quincy Colton successfully started to use it in all his "Colton Dental Association" clinics, that he had just established in New Haven and New York City. Over the following three years, Colton and his associates successfully administered nitrous oxide to more than 25,000 patients. Today, nitrous oxide is used in dentistry as an anxiolytic, as an adjunct to local anaesthetic.

However, nitrous oxide was not found to be a strong enough anaesthetic for use in major surgery in hospital settings. Being a stronger and more potent anaesthetic, sulfuric ether was instead demonstrated and accepted for use in October 1846, along with chloroform in 1847. When Joseph Thomas Clover invented the "gas-ether inhaler" in 1876, it however became a common practice at hospitals to initiate all anaesthetic treatments with a mild flow of nitrous oxide, and then gradually increase the anaesthesia with the stronger ether/chloroform. Clover's gas-ether inhaler was designed to supply the patient with nitrous oxide and ether at the same time, with the exact mixture being controlled by the operator of the device. It remained in use by many hospitals until the 1930s. Although hospitals today are using a more advanced anaesthetic machine, these machines still use the same principle launched with Clover's gas-ether inhaler, to initiate the anaesthesia with nitrous oxide, before the administration of a more powerful anaesthetic.

As a Patent Medicine

Colton's popularization of nitrous oxide led to its adoption by a number of less than reputable quacksalvers, who touted it as a cure for consumption, scrofula, catarrh, and other diseases of the blood, throat, and lungs. Nitrous oxide treatment was administered and licensed as a patent medicine by the likes of C. L. Blood and Jerome Harris in Boston and Charles E. Barney of Chicago.

Legality

In the United States, possession of nitrous oxide is legal under federal law and is not subject to DEA purview. It is, however, regulated by the Food and Drug Administration under the Food Drug and Cosmetics Act; prosecution is possible under its "misbranding" clauses, prohibiting the sale or distribution of nitrous oxide for the purpose of human consumption.

Many states have laws regulating the possession, sale, and distribution of nitrous oxide. Such laws usually ban distribution to minors or limit the amount of nitrous oxide that may be sold without special license. For example, in the state of California, possession for recreational use is prohibited and qualifies as a misdemeanour.

In New Zealand, the Ministry of Health has warned that nitrous oxide is a prescription medicine, and its sale or possession without a prescription is an offence under the Medicines Act. This statement would seemingly prohibit all non-medicinal uses of the chemical, though it is implied that only recreational use will be legally targeted.

In India, for general anaesthesia purposes, nitrous oxide is available as Nitrous Oxide IP. India's gas cylinder rules (1985) permit the transfer of gas from one cylinder to another for breathing purposes. This law benefits remote hospitals, which would otherwise suffer as a result of India's geographic immensity. Nitrous Oxide IP is transferred from bulk cylinders (17,000 litres [600 cu ft] capacity gas) to smaller pin-indexed valve cylinders (1,800 litres [64 cu ft] of gas), which are then connected to the yoke assembly of Boyle's machines. Because India's Food & Drug Authority (FDA-India) rules state that transferring a drug from one container to another (refilling) is equivalent to manufacturing, anyone found doing so must possess a drug manufacturing license.

Nitrogen Trifluoride

Nitrogen trifluoride is the inorganic compound with the formula NF_3. This nitrogen-fluorine compound is a colorless, odorless, nonflammable gas. It finds increasing use as an etchant in microelectronics.

Applications

Nitrogen trifluoride is used in the plasma etching of silicon wafers. Today nitrogen trifluoride is predominantly employed in the cleaning of the PECVD chambers in the high-volume production of liquid-crystal displays and silicon-based thin-film solar cells. In these applications NF_3 is initially broken down *in situ* by a plasma. The resulting fluorine atoms are the active cleaning agents that attack the polysilicon, silicon nitride and silicon oxide. Nitrogen trifluoride can be used as well with tungsten silicide, and tungsten produced by CVD. NF_3 has been considered as an environmentally preferable substitute for sulfur hexafluoride or perfluorocarbons such as hexafluoroethane. The process utilization of the chemicals applied in plasma processes is typically below 20%. Therefore some of the PFCs and also some of the NF_3 always escape into the atmosphere. Modern gas abatement systems can decrease such emissions.

Elemental fluorine has been introduced as an environmentally friendly replacement for nitrogen trifluoride in the manufacture of flat-panel displays and thin-film solar cells.

Nitrogen trifluoride is also used in hydrogen fluoride and deuterium fluoride lasers, which are types of chemical lasers. It is preferred to fluorine gas due to its convenient handling properties, reflecting its considerable stability.

It is compatible with steel and Monel, as well as several plastics.

Synthesis and Reactivity

Nitrogen trifluoride is a rare example of a binary fluoride that can be prepared directly from the elements only at very uncommon conditions, such as electric discharge. After first attempting the synthesis in 1903, Otto Ruff prepared nitrogen trifluoride by the electrolysis of a molten mixture of ammonium fluoride and hydrogen fluoride. It proved to be far less reactive than the other nitrogen trihalides nitrogen trichloride, nitrogen tribromide and nitrogen triiodide, all of which are explosive. Alone among the nitrogen trihalides it has a negative enthalpy of formation. Today, it is prepared both by direct reaction of ammonia and fluorine and by a variation of Ruff's method. It is supplied in pressurized cylinders.

Reactions

NF_3 is slightly soluble in water without undergoing chemical reaction. It is nonbasic with a low dipole moment of 0.2340 D. By contrast, ammonia is basic and highly polar (1.47 D). This difference arises from the fluorine atoms acting as electron withdrawing groups, attracting essentially all of the lone pair electrons on the nitrogen atom. NF_3 is a potent yet sluggish oxidizer.

It oxidizes hydrogen chloride to chlorine:

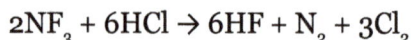

$$2NF_3 + 6HCl \rightarrow 6HF + N_2 + 3Cl_2$$

It converts to tetrafluorohydrazine upon contact with metals, but only at high temperatures:

$$2NF_3 + Cu \rightarrow N_2F_4 + CuF_2$$

NF_3 reacts with fluorine and antimony pentafluoride to give the tetrafluoroammonium salt:

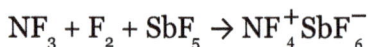

$$NF_3 + F_2 + SbF_5 \rightarrow NF_4^+SbF_6^-$$

Greenhouse Gas

NF3 is a greenhouse gas, with a global warming potential (GWP) 17,200 times greater than that of CO2 when compared over a 100-year period. Its GWP place it second only to SF6 in the group of Kyoto-recognised greenhouse gases, and NF3 was included in that grouping with effect from 2013 and the commencement of the second commitment period of the Kyoto Protocol. It has an estimated atmospheric lifetime of 740 years, although other work suggests a slightly shorter lifetime of 550 years (and a corresponding GWP of 16,800).

Although NF3 has a high GWP, for a long time its radiative forcing in the Earth's atmosphere has been assumed to be small, spuriously presuming that only small quantities are released into the atmosphere. Industrial applications of NF3 routinely break it down, while in the past previously used regulated compounds such as SF6 and PFCs were often released. Research has questioned the previous assumptions. High-volume applications such as DRAM computer memory production, the manufacturing of flat panel displays and the large-scale production of thin-film solar cells.

Since 1992, when less than 100 tons were produced, production has grown to an estimated 4000 tons in 2007 and is projected to increase significantly. World production of NF_3 is expected to reach 8000 tons a year by 2010. By far the world's largest producer of NF3 is the US industrial gas and chemical company Air Products & Chemicals. An estimated 2% of produced NF3 is released into the atmosphere. Robson projected that the maximum atmospheric concentration is less than 0.16 parts per trillion (ppt) by volume, which will provide less than 0.001 Wm^{-2} of IR forcing. The mean global tropospheric concentration of NF_3 has risen from about 0.02 ppt (parts per trillion, dry air mole fraction) in 1980, to 0.86 ppt in 2011, with a rate of increase of 0.095 ppt yr^{-1}, or about 11% per year, and an interhemispheric gradient that is consistent with emissions occurring overwhelmingly in the Northern Hemisphere, as expected. This rise rate in 2011 corresponds to about 1200 metric tons/y NF_3 emissions globally, or about 10% of the NF_3 global production estimates. This is a significantly higher percentage than has been estimated by industry, and thus strengthens the case for inventorying NF_3 production and for regulating its emissions. One study co-authored by industry representatives suggests that the contribution of the NF_3 emissions to the overall greenhouse gas budget of thin-film Si-solar cell manufacturing is overestimated. Instead, the contribution of the nitrogen trifluoride to the CO_2-budget of thin film solar cell production is compensated already within a few months by the CO_2 saving potential of the PV technology. However, unlike CO_2, NF_3 does not interact with the carbon cycle and therefore has no natural means of sequestration once emitted.

The UNFCCC, within the context of the Kyoto Protocol, decided to include nitrogen trifluoride in the second Kyoto Protocol compliance period, which begins in 2012 and ends in either 2017 or 2020. Following suit, the WBCSD/WRI GHG Protocol is amending all of its standards (corporate, product and Scope 3) to also cover NF_3.

Safety

Skin contact with NF3 is not hazardous, and it is a relatively minor irritant to mucous membranes and eyes. It is a pulmonary irritant with a toxicity comparable with nitrogen oxides, and over-exposure via inhalation causes the conversion of hemoglobin in blood to methemoglobin, which can lead to the condition methemoglobinemia. The National Institute for Occupational Safety and Health (NIOSH) specifies that the concentration that is immediately dangerous to life or health (IDLH value) is 1,000 ppm.

References

- Donald G. Kaufman; Cecilia M. Franz (1996). Biosphere 2000: protecting our global environment. Kendall/Hunt Pub. Co. ISBN 978-0-7872-0460-0. Retrieved 11 October 2011.

- Ocean Acidification: A National Strategy to Meet the Challenges of a Changing Ocean. Washington, DC: National Academies Press. doi:10.17226/12904. ISBN 978-0-309-15359-1.

- Greenwood, Norman N.; Earnshaw, Alan (1997). Chemistry of the Elements (2nd ed.). Butterworth-Heinemann. ISBN 0-08-037941-9.

- Pierantozzi, Ronald (2001). "Carbon Dioxide". Kirk-Othmer Encyclopedia of Chemical Technology. Kirk-Othmer Encyclopedia of Chemical Technology. Wiley. doi:10.1002/0471238961.0301180216090518.a01.pub2. ISBN 0-471-23896-1.

- Strassburger, Julius (1969). Blast Furnace Theory and Practice. New York: American Institute of Mining, Metallurgical, and Petroleum Engineers. ISBN 0-677-10420-0.

- Martini, M. (1997). "CO_2 emissions in volcanic areas: case histories and hazards". In Raschi, A.; Miglietta, F.; Tognetti, R.; van Gardingen, P.R. Plant responses to elevated CO_2: Evidence from natural springs. Cambridge: Cambridge University Press. pp. 69–86. ISBN 0-521-58203-2.

- Hensher, David A. & Button, Kenneth J. (2003). Handbook of transport and the environment. Emerald Group Publishing. p. 168. ISBN 0-08-044103-3.

- Drysdale, Dougal (2008). "Physics and Chemistry of Fire". In Cote, Arthur E. Fire Protection Handbook. 1 (20th ed.). Quincy, MA: National Fire Protection Association. pp. 2–18. ISBN 978-0-87765-758-3.

- Miller, G. Tyler (2007). Sustaining the Earth: An Integrated Approach. U.S.A.: Thomson Advantage Books, ISBN 0534496725.

- Miller, Ron; Hartmann, William K. (2005). The Grand Tour: A Traveler's Guide to the Solar System (3rd ed.). Thailand: Workman Publishing. pp. 172–73. ISBN 0-7611-3547-2.

- Jørgensen, Uffe G. (1997), "Cool Star Models", in van Dishoeck, Ewine F., Molecules in Astrophysics: Probes and Processes, International Astronomical Union Symposia. Molecules in Astrophysics: Probes and Processes, 178, Springer Science & Business Media, p. 446, ISBN 079234538X.

- Brown, Theodore L.; LeMay, H. Eugene, Jr.; Bursten, Bruce E.; Burdge, Julia R. (2003) [1977]. "22". In Nicole Folchetti. Chemistry: The Central Science (9th ed.). Pearson Education. pp. 882–883. ISBN 0-13-066997-0.

- Solomons, T.W. Graham & Fryhle, Craig B. (2008). "Chapter 8 Alkenes and Alkynes – Part II: Addition Reactions and Synthesis". Organic Chemistry, 9th Edition. Wiley. p. 344. ISBN 978-0-470-16982-7.

- Housecroft, C. E.; Sharpe, A. G. (2004). Inorganic Chemistry (2nd ed.). Prentice Hall. p. 439. ISBN 978-0130399137.

- Sjöström, Eero (1993). Wood Chemistry: Fundamentals and Applications. San Diego, CA: Academic Press, Inc. ISBN 0-12-647481-8.

- Jongen, W (2005). Improving the Safety of Fresh Fruit and Vegetables. Boca Raton: Woodhead Publishing Ltd. ISBN 1-85573-956-9.

- Sneader W (2005). Drug Discovery –A History. (Part 1: Legacy of the past, chapter 8: systematic medicine, pp. 74–87). John Wiley and Sons. ISBN 978-0-471-89980-8. Retrieved 21 April 2010.

- Housecroft, Catherine E. & Sharpe, Alan G. (2008). "Chapter 15: The group 15 elements". Inorganic Chemistry (3rd ed.). Pearson. p. 464. ISBN 978-0-13-175553-6.

3

Effects of Greenhouse Gas

Greenhouse gases and aerosols enter the earth's higher atmospheric levels and allow solar radiation to penetrate the earth's surface. These gases also re-emit solar energy onto the land and oceans, causing a rise in temperature levels. Some of the topics discussed in this chapter are radiative forcing and the greenhouse effect. This chapter elucidates the crucial theories and principles of greenhouse gas emissions.

Greenhouse Effect

The greenhouse effect is the process by which radiation from a planet's atmosphere warms the planet's surface to a temperature above what it would be without its atmosphere.

If a planet's atmosphere contains radiatively active gases (i.e., greenhouse gases) the atmosphere will radiate energy in all directions. Part of this radiation is directed towards the surface, warming it. The downward component of this radiation – that is, the strength of the greenhouse effect – will depend on the atmosphere's temperature and on the amount of greenhouse gases that the atmosphere contains.

On Earth, the atmosphere is warmed by absorption of infrared thermal radiation from the underlying surface, absorption of shorter wavelength radiant energy from the sun, and convective heat fluxes from the surface. Greenhouse gases in the atmosphere radiate energy, some of which is directed to the surface and lower atmosphere. The mechanism that produces this difference between the actual surface temperature and the effective temperature is due to the atmosphere and is known as the greenhouse effect.

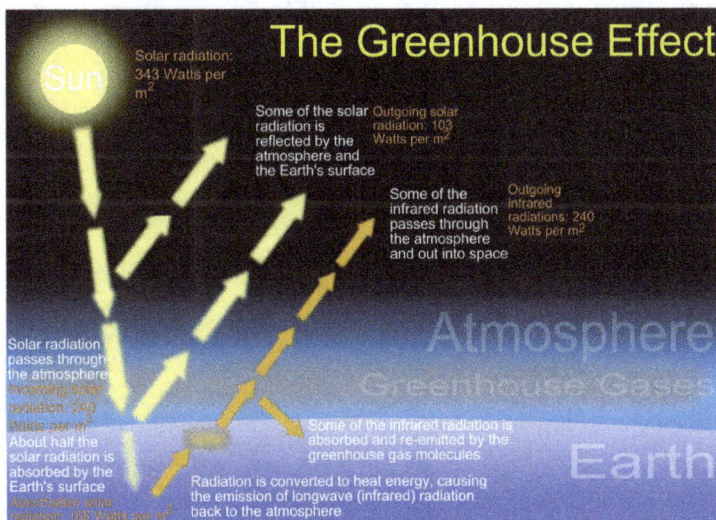

Another diagram of the greenhouse effect

Earth's natural greenhouse effect is critical to supporting life. Human activities, primarily the burning of fossil fuels and clearing of forests, have intensified the natural greenhouse effect, causing global warming.

The mechanism is named after a faulty analogy with the effect of solar radiation passing through glass and warming a greenhouse. The way a greenhouse retains heat is fundamentally different, as a greenhouse works by reducing airflow and retaining warm air inside the structure.

History

The existence of the greenhouse effect was argued for by Joseph Fourier in 1824. The argument and the evidence was further strengthened by Claude Pouillet in 1827 and 1838, and reasoned from experimental observations by John Tyndall in 1859. The effect was more fully quantified by Svante Arrhenius in 1896. However, the term "greenhouse" was not used to refer to this effect by any of these scientists; the term was first used in this way by Nils Gustaf Ekholm in 1901.

In 1917 Alexander Graham Bell wrote "[The unchecked burning of fossil fuels] would have a sort of greenhouse effect", and "The net result is the greenhouse becomes a sort of hot-house." Bell went on to also advocate the use of alternate energy sources, such as solar energy.

Mechanism

Earth receives energy from the Sun in the form of ultraviolet, visible, and near-infrared radiation. Of the total amount of solar energy available at the top of the atmosphere, about 26% is reflected to space by the atmosphere and clouds and 19% is absorbed by the atmosphere and clouds. Most of the remaining energy is absorbed at the surface of Earth. Because the Earth's surface is colder than the photosphere of the Sun, it radiates at wavelengths that are much longer than the wavelengths that were absorbed. Most of this thermal radiation is absorbed by the atmosphere, thereby warming it. In addition to the absorption of solar and thermal radiation, the atmosphere further gains heat by sensible and latent heat fluxes from the surface. The atmosphere radiates energy both upwards and downwards; the part radiated downwards is absorbed by the surface of Earth. This leads to a higher equilibrium temperature than if the atmosphere were absent.

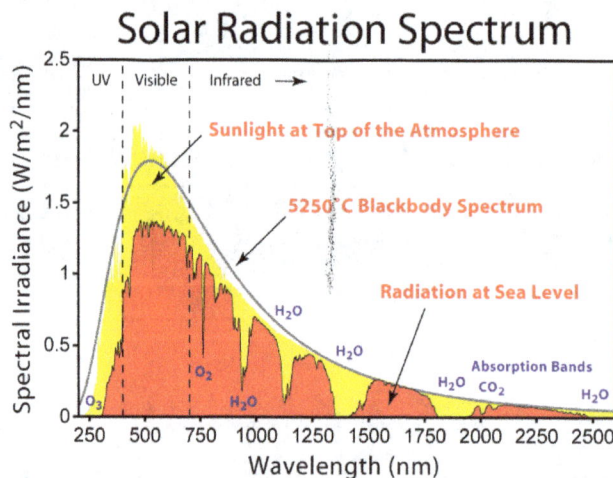

The solar radiation spectrum for direct light at both the top of Earth's atmosphere and at sea level

An ideal thermally conductive blackbody at the same distance from the Sun as Earth would have a temperature of about 5.3 °C. However, because Earth reflects about 30% of the incoming sunlight, this idealized planet's effective temperature (the temperature of a blackbody that would emit the same amount of radiation) would be about −18 °C. The surface temperature of this hypothetical planet is 33 °C below Earth's actual surface temperature of approximately 14 °C.

The basic mechanism can be qualified in a number of ways, none of which affect the fundamental process. The atmosphere near the surface is largely opaque to thermal radiation (with important exceptions for "window" bands), and most heat loss from the surface is by sensible heat and latent heat transport. Radiative energy losses become increasingly important higher in the atmosphere, largely because of the decreasing concentration of water vapor, an important greenhouse gas. It is more realistic to think of the greenhouse effect as applying to a "surface" in the mid-troposphere, which is effectively coupled to the surface by a lapse rate. The simple picture also assumes a steady state, but in the real world there are variations due to the diurnal cycle as well as the seasonal cycle and weather disturbances. Solar heating only applies during daytime. During the night, the atmosphere cools somewhat, but not greatly, because its emissivity is low. Diurnal temperature changes decrease with height in the atmosphere.

Within the region where radiative effects are important, the description given by the idealized greenhouse model becomes realistic. Earth's surface, warmed to a temperature around 255 K, radiates long-wavelength, infrared heat in the range of 4–100 µm. At these wavelengths, greenhouse gases that were largely transparent to incoming solar radiation are more absorbent. Each layer of atmosphere with greenhouses gases absorbs some of the heat being radiated upwards from lower layers. It reradiates in all directions, both upwards and downwards; in equilibrium (by definition) the same amount as it has absorbed. This results in more warmth below. Increasing the concentration of the gases increases the amount of absorption and reradiation, and thereby further warms the layers and ultimately the surface below.

Greenhouse gases—including most diatomic gases with two different atoms (such as carbon monoxide, CO) and all gases with three or more atoms—are able to absorb and emit infrared radiation. Though more than 99% of the dry atmosphere is IR transparent (because the main constituents—N_2, O_2, and Ar—are not able to directly absorb or emit infrared radiation), intermolecular collisions cause the energy absorbed and emitted by the greenhouse gases to be shared with the other, non-IR-active, gases.

Greenhouse Gases

Atmospheric gases only absorb some wavelengths of energy but are transparent to others. The absorption patterns of water vapor (blue peaks) and carbon dioxide (pink peaks) overlap in some wavelengths. Carbon dioxide is not as strong a greenhouse gas as water vapor, but it absorbs energy in wavelengths (12-15 micrometers) that water vapor does not, partially closing the "window" through which heat radiated by the surface would normally escape to space. (Illustration NASA, Robert Rohde)

By their percentage contribution to the greenhouse effect on Earth the four major gases are:

- water vapor, 36–70%

- carbon dioxide, 9–26%

- methane, 4–9%

- ozone, 3–7%

It is not physically realistic to assign a specific percentage to each gas because the absorption and emission bands of the gases overlap (hence the ranges given above). The major non-gas contributor to Earth's greenhouse effect, clouds, also absorb and emit infrared radiation and thus have an effect on the radiative properties of the atmosphere.

Role in Climate Change

Strengthening of the greenhouse effect through human activities is known as the enhanced (or anthropogenic) greenhouse effect. This increase in radiative forcing from human activity is attributable mainly to increased atmospheric carbon dioxide levels. According to the latest Assessment Report from the Intergovernmental Panel on Climate Change, *"atmospheric concentrations of carbon dioxide, methane and nitrous oxide are unprecedented in at least the last 800,000 years. Their effects, together with those of other anthropogenic drivers, have been detected throughout the climate system and are extremely likely to have been the dominant cause of the observed warming since the mid-20th century"*.

CO_2 is produced by fossil fuel burning and other activities such as cement production and tropical deforestation. Measurements of CO_2 from the Mauna Loa observatory show that concentrations have increased from about 313 parts per million (ppm) in 1960 to about 389 ppm in 2010. It reached the 400 ppm milestone on May 9, 2013. The current observed amount of CO_2 exceeds the geological record maxima (~300 ppm) from ice core data. The effect of combustion-produced carbon dioxide on the global climate, a special case of the greenhouse effect first described in 1896 by Svante Arrhenius, has also been called the Callendar effect.

Over the past 800,000 years, ice core data shows that carbon dioxide has varied from values as low as 180 ppm to the pre-industrial level of 270 ppm. Paleoclimatologists consider variations in carbon dioxide concentration to be a fundamental factor influencing climate variations over this time scale.

Real Greenhouses

The "greenhouse effect" of the atmosphere is named by analogy to greenhouses which become warmer in sunlight. The explanation given in most sources for the warmer temperature in an actual greenhouse is that incident solar radiation in the visible, long-wavelength ultraviolet, and short-wavelength infrared range of the spectrum passes through the glass roof and walls and is absorbed by the floor, earth, and contents, which become warmer and re-emit the energy as longer-wavelength infrared radiation. Glass and other materials used for greenhouse walls do not transmit infrared radiation, so the infrared cannot escape via radiative transfer. As the structure is not open to the atmosphere, heat also cannot escape via convection, so the temperature inside the greenhouse rises. The greenhouse effect, due to infrared-opaque "greenhouse gases" including carbon dioxide and methane instead of glass, also affects Earth as a whole; there is no convective cooling because no significant amount of air escapes from Earth.

A modern Greenhouse in RHS Wisley

However the mechanism by which the atmosphere retains heat—the "greenhouse effect"—is different; a greenhouse is not primarily warmed by the "greenhouse effect". A greenhouse works primarily by allowing sunlight to warm surfaces inside the structure, but then preventing absorbed heat from leaving the structure through convection. The "greenhouse effect" heats Earth because greenhouse gases absorb outgoing radiative energy, heating the atmosphere which then emits radiative energy with some of it going back towards Earth.

A greenhouse is built of any material that passes sunlight, usually glass, or plastic. It mainly warms up because the sun warms the ground and contents inside, which then warms the air in the greenhouse. The air continues to heat up because it is confined within the greenhouse, unlike the environment outside the greenhouse where warm air near the surface rises and mixes with cooler air aloft. This can be demonstrated by opening a small window near the roof of a greenhouse: the temperature will drop considerably. It was demonstrated experimentally (R. W. Wood, 1909) that a "greenhouse" with a cover of rock salt (which is transparent to infrared) heats up an enclosure similarly to one with a glass cover. Thus greenhouses work primarily by preventing convective cooling.

More recent quantitative studies suggest that the effect of infrared radiative cooling is not negligibly small, and may have economic implications in a heated greenhouse. Analysis of issues of near-infrared radiation in a greenhouse with screens of a high coefficient of reflection concluded that installation of such screens reduced heat demand by about 8%, and application of dyes to transparent surfaces was suggested. Composite less-reflective glass, or less effective but cheaper anti-reflective coated simple glass, also produced savings.

Bodies other than Earth

In the Solar System, there also greenhouse effects on Mars, Venus, and Titan. The greenhouse effect on Venus is particularly large because its dense atmosphere consisting mainly of carbon dioxide. Titan has an anti-greenhouse effect, in that its atmosphere absorbs solar radiation but is relatively transparent to infrared radiation. Pluto is also colder than would be expected, because evaporation of nitrogen cools it.

A runaway greenhouse effect occurs if positive feedbacks lead to the evaporation of all greenhouse gases into the atmosphere. A runaway greenhouse effect involving carbon dioxide and water vapor is thought to have occurred on Venus.

Radiative Forcing

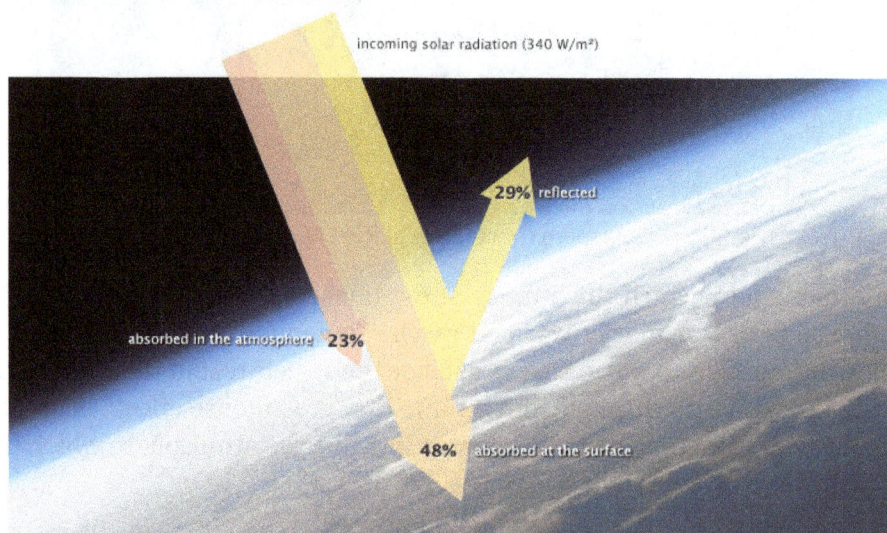

Incoming solar radiation

Radiative forcing or climate forcing is defined as the difference of insolation (sunlight) absorbed by the Earth and energy radiated back to space. Typically, radiative forcing is quantified at the tropopause in units of watts per square meter of the Earth's surface. A positive forcing (more incoming energy) warms the system, while negative forcing (more outgoing energy) cools it. Causes of radiative forcing include changes in insolation and the concentrations of radiatively active gases, commonly known as greenhouse gases and aerosols.

Radiation Balance

Atmospheric gases only absorb some wavelengths of energy but are transparent to others. The absorption patterns of water vapor (blue peaks) and carbon dioxide (pink peaks) overlap in some wavelengths. Carbon dioxide is not as strong a greenhouse gas as water vapor, but it absorbs energy in wavelengths (12-15 micrometers) that water vapor does not, partially closing the "window" through which heat radiated by the surface would normally escape to space. (Illustration NASA, Robert Rohde)

Almost all of the energy that affects Earth's climate is received as radiant energy from the Sun. The planet and its atmosphere absorb and reflect some of the energy, while long-wave energy is radiated back into space. The balance between absorbed and radiated energy determines the average global temperature. Because the atmosphere absorbs some of the re-radiated long-wave energy, the planet is warmer than it would be in the absence of the atmosphere: see greenhouse effect.

The radiation balance is altered by such factors as the intensity of solar energy, reflectivity of clouds or gases, absorption by various greenhouse gases or surfaces and heat emission by various materials. Any such alteration is a radiative forcing, and changes the balance. This happens continuously as sunlight hits the surface, clouds and aerosols form, the concentrations of atmospheric gases vary and seasons alter the ground cover.

IPCC Usage

Radiative forcing components

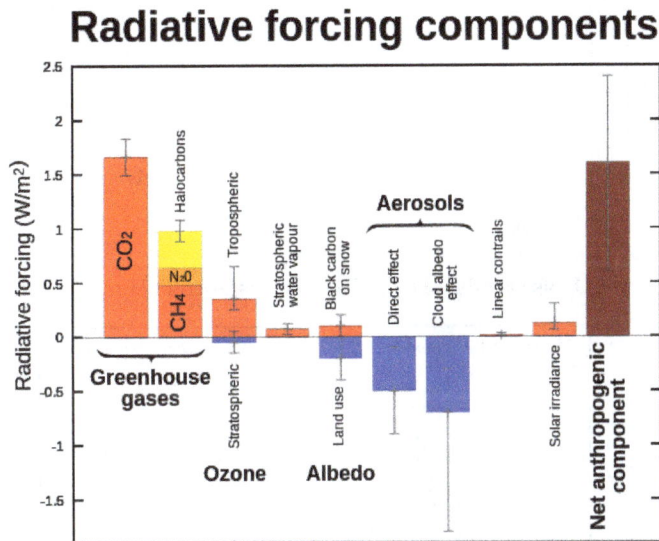

Radiative forcings, IPCC 2007.

The Intergovernmental Panel on Climate Change (IPCC) AR4 report defines radiative forcings as:

"Radiative forcing is a measure of the influence a factor has in altering the balance of incoming and outgoing energy in the Earth-atmosphere system and is an index of the importance of the factor as a potential climate change mechanism. In this report radiative forcing values are for changes relative to preindustrial conditions defined at 1750 and are expressed in Watts per square meter (W/m²)."

In simple terms, radiative forcing is "...the rate of energy change per unit area of the globe as measured at the top of the atmosphere." In the context of climate change, the term "forcing" is restricted to changes in the radiation balance of the surface-troposphere system imposed by external factors, with no changes in stratospheric dynamics, no surface and tropospheric feedbacks in operation (i.e., no secondary effects induced because of changes in tropospheric motions or its thermodynamic state), and no dynamically induced changes in the amount and distribution of atmospheric water (vapour, liquid, and solid forms).

Climate Sensitivity

Radiative forcing can be used to estimate a subsequent change in equilibrium surface temperature

(ΔT_s) arising from that forcing via the equation:

$$\Delta T_s = \lambda \Delta F$$

where λ is the climate sensitivity, usually with units in K/(W/m²), and ΔF is the radiative forcing. A typical value of λ is 0.8 K/(W/m²), which gives a warming of 3K for doubling of CO_2.

Example Calculations

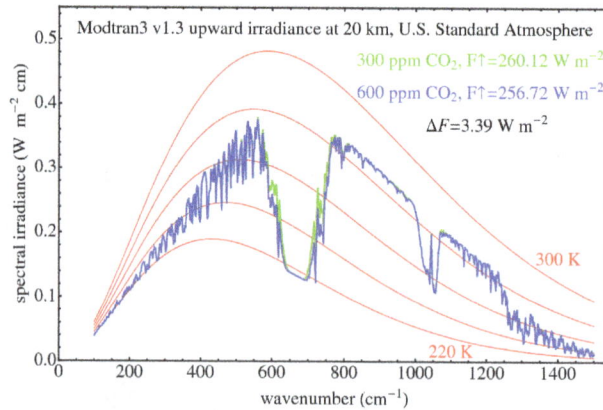

Radiative forcing for doubling CO_2, as calculated by radiative transfer code Modtran. Red lines are Planck curves.

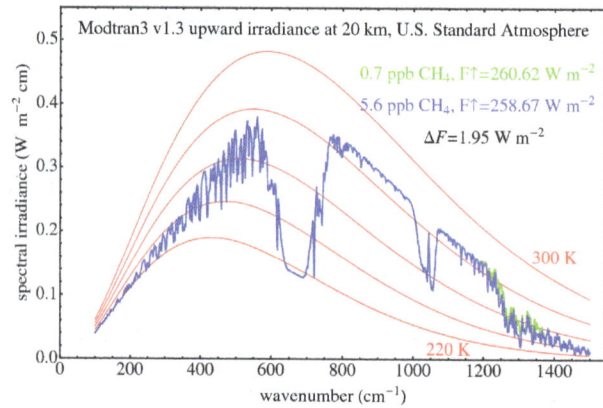

Radiative forcing for eight times increase of CH_4, as calculated by radiative transfer code Modtran.

Solar Forcing

Radiative forcing (measured in Watts per square meter) can be estimated in different ways for different components. For solar irradiance (*i.e.,* "solar forcing"), the radiative forcing is simply the change in the average amount of solar energy absorbed per square meter of the Earth's area. Since the Earth's cross-sectional area exposed to the Sun (πr^2) is equal to 1/4 of the surface area of the Earth ($4\pi r^2$), the solar input per unit area is one quarter the change in solar intensity. This must be multiplied by the fraction of incident sunlight that is absorbed, F=(1-R), where R is the reflectivity (albedo), of the Earth. The albedo is approximately 0.3, so F is approximately equal to 0.7. Thus, the solar forcing is the change in the solar intensity divided by 4 and multiplied by 0.7.

Likewise, a change in albedo will produce a solar forcing equal to the change in albedo divided by 4 multiplied by the solar constant.

Forcing due to Atmospheric Gas

For a greenhouse gas, such as carbon dioxide, radiative transfer codes that examine each spectral line for atmospheric conditions can be used to calculate the change ΔF as a function of changing concentration. These calculations can often be simplified into an algebraic formulation that is specific to that gas.

For instance, the simplified first-order approximation expression for carbon dioxide is:

$$\Delta F = 5.35 \times \ln \frac{C}{C_0} \, \text{W m}^{-2}$$

where C is the CO_2 concentration in parts per million by volume and C_0 is the reference concentration. The relationship between carbon dioxide and radiative forcing is logarithmic and thus increased concentrations have a progressively smaller warming effect.

A different formula applies for other greenhouse gases such as methane and N_2O (square-root dependence) or CFCs (linear), with coefficients that can be found *e.g.* in the IPCC reports.

Related Measures

Radiative forcing is a useful way to compare different causes of perturbations in a climate system. Other possible tools can be constructed for the same purpose: for example Shine *et al.* say "... recent experiments indicate that for changes in absorbing aerosols and ozone, the predictive ability of radiative forcing is much worse... we propose an alternative, the 'adjusted troposphere and stratosphere forcing'. We present GCM calculations showing that it is a significantly more reliable predictor of this GCM's surface temperature change than radiative forcing. It is a candidate to supplement radiative forcing as a metric for comparing different mechanisms...". In this quote, GCM stands for "global circulation model", and the word "predictive" does not refer to the ability of GCMs to forecast climate change. Instead, it refers to the ability of the alternative tool proposed by the authors to help explain the system response.

History

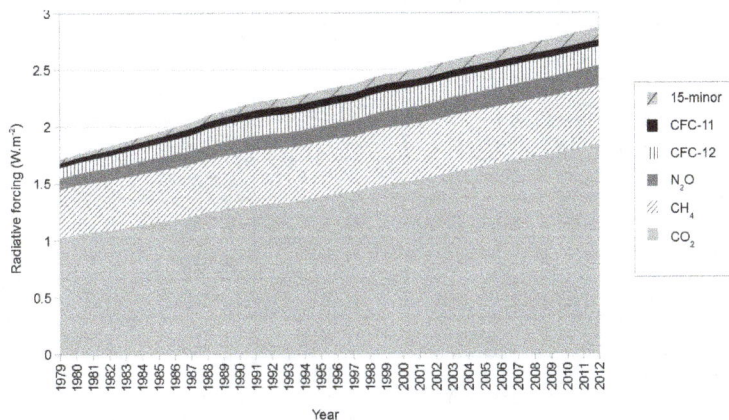

Changes in radiative forcing of long-lived greenhouse gases between 1979 and 2012.

The table below (derived from atmospheric radiative transfer models) shows changes in radiative forcing between 1979 and 2013. The table includes the contribution to radiative forcing from carbon dioxide (CO_2), methane (CH4), nitrous oxide (N2O); chlorofluorocarbons (CFCs) 12 and 11;

and fifteen other minor, long-lived, halogenated gases. The table includes the contribution to radiative forcing of long-lived greenhouse gases. It does not include other forcings, such as aerosols and changes in solar activity.

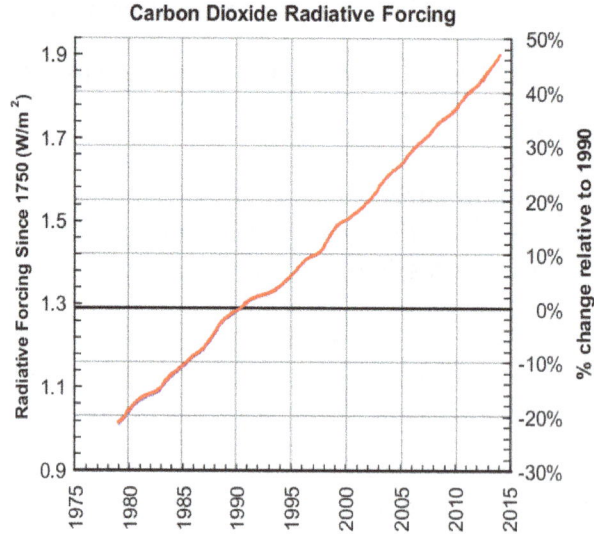

Carbon Dioxide Radiative Forcing

Radiative forcing, relative to 1750, due to carbon dioxide alone since 1979. The percent change from January 1, 1990 is shown on the right axis.

Global radiative forcing (relative to 1750, in), CO_2-equivalent mixing ratio, and the Annual Greenhouse Gas Index (AGGI) between 1979-2014										
Year	CO_2	CH4	N2O	CFC-12	CFC-11	15-minor	Total	CO_2-eq ppm	AGGI 1990 = 1	AGGI % change
1979	1.027	0.419	0.104	0.092	0.039	0.031	1.712	383	0.786	
1980	1.058	0.426	0.104	0.097	0.042	0.034	1.761	386	0.808	2.8
1981	1.077	0.433	0.107	0.102	0.044	0.036	1.799	389	0.826	2.2
1982	1.089	0.440	0.111	0.108	0.046	0.038	1.831	391	0.841	1.8
1983	1.115	0.443	0.113	0.113	0.048	0.041	1.873	395	0.860	2.2
1984	1.140	0.446	0.116	0.118	0.050	0.044	1.913	397	0.878	2.2
1985	1.162	0.451	0.118	0.123	0.053	0.047	1.953	401	0.897	2.1
1986	1.184	0.456	0.122	0.129	0.056	0.049	1.996	404	0.916	2.2
1987	1.211	0.460	0.120	0.135	0.059	0.053	2.039	407	0.936	2.2
1988	1.250	0.464	0.123	0.143	0.062	0.057	2.099	412	0.964	3.0
1989	1.274	0.468	0.126	0.149	0.064	0.061	2.144	415	0.984	2.1
1990	1.293	0.472	0.129	0.154	0.065	0.065	2.178	418	1.000	1.6
1991	1.313	0.476	0.131	0.158	0.067	0.069	2.213	420	1.016	1.6
1992	1.324	0.480	0.133	0.162	0.067	0.072	2.238	422	1.027	1.1
1993	1.334	0.481	0.134	0.164	0.068	0.074	2.254	424	1.035	0.7
1994	1.356	0.483	0.134	0.166	0.068	0.075	2.282	426	1.048	1.3
1995	1.383	0.485	0.136	0.168	0.067	0.077	2.317	429	1.064	1.5
1996	1.410	0.486	0.139	0.169	0.067	0.078	2.350	431	1.079	1.4
1997	1.426	0.487	0.142	0.171	0.067	0.079	2.372	433	1.089	0.9

1998	1.465	0.491	0.145	0.172	0.067	0.080	2.419	437	1.111	2.0
1999	1.495	0.494	0.148	0.173	0.066	0.082	2.458	440	1.128	1.6
2000	1.513	0.494	0.151	0.173	0.066	0.083	2.481	442	1.139	0.9
2001	1.535	0.494	0.153	0.174	0.065	0.085	2.506	444	1.150	1.0
2002	1.564	0.494	0.156	0.174	0.065	0.087	2.539	447	1.166	1.3
2003	1.601	0.496	0.158	0.174	0.064	0.088	2.580	450	1.185	1.6
2004	1.627	0.496	0.160	0.174	0.063	0.090	2.610	453	1.198	1.1
2005	1.655	0.495	0.162	0.173	0.063	0.092	2.640	455	1.212	1.2
2006	1.685	0.495	0.165	0.173	0.062	0.095	2.675	458	1.228	1.3
2007	1.710	0.498	0.167	0.172	0.062	0.097	2.706	461	1.242	1.1
2008	1.739	0.500	0.170	0.171	0.061	0.100	2.742	464	1.259	1.3
2009	1.760	0.502	0.172	0.171	0.061	0.103	2.768	466	1.271	1.0
2010	1.791	0.504	0.174	0.170	0.060	0.106	2.805	470	1.288	1.3
2011	1.818	0.505	0.178	0.169	0.060	0.109	2.838	473	1.303	1.2
2012	1.846	0.507	0.181	0.168	0.059	0.111	2.873	476	1.319	1.2
2013	1.884	0.509	0.184	0.167	0.059	0.114	2.916	479	1.338	1.5
2014	1.909	0.500	0.187	0.166	0.058	0.116	2.936	481	1.356	1.6

The table shows that CO_2 dominates the total forcing, with methane and chlorofluorocarbons (CFC) becoming relatively smaller contributors to the total forcing over time. The five major greenhouse gases account for about 96% of the direct radiative forcing by long-lived greenhouse gas increases since 1750. The remaining 4% is contributed by the 15 minor halogenated gases.

The table also includes an "Annual Greenhouse Gas Index" (AGGI), which is defined as the ratio of the total direct radiative forcing due to long-lived greenhouse gases for any year for which adequate global measurements exist to that which was present in 1990. 1990 was chosen because it is the baseline year for the Kyoto Protocol. This index is a measure of the inter-annual changes in conditions that affect carbon dioxide emission and uptake, methane and nitrous oxide sources and sinks, the decline in the atmospheric abundance of ozone-depleting chemicals related to the Montreal Protocol. and the increase in their substitutes (hydrogenated CFCs (HCFCs) and hydrofluorocarbons (HFC). Most of this increase is related to CO_2. For 2013, the AGGI was 1.34 (representing an increase in total direct radiative forcing of 34% since 1990). The increase in CO_2 forcing alone since 1990 was about 46%. The decline in CFCs considerably tempered the increase in net radiative forcing.

References

- Vaclav Smil (2003). The Earth's Biosphere: Evolution, Dynamics, and Change. MIT Press. p. 107. ISBN 978-0-262-69298-4.

- Grosvenor, Edwin S. and Morgan Wesson. Alexander Graham Bell: The Life and Times of the Man Who Invented the Telephone. New York: Harry N. Abrahms, Inc., 1997, p. 274, ISBN 0-8109-4005-1.

- A Dictionary of Chemistry (6 ed.), edited by John Daintith, Publisher: Oxford University Press, 2008, ISBN 9780199204632

- Brian Shmaefsky (2004). Favorite demonstrations for college science: an NSTA Press journals collection. NSTA Press. p. 57. ISBN 978-0-87355-242-4.

- Energy effects during using the glass with different properties in a heated greenhouse, sławomir kurpaska,technical sciences 17(4), 2014, 351–360

- "Titan: Greenhouse and Anti-greenhouse :: Astrobiology Magazine - earth science - evolution distribution Origin of life universe - life beyond :: Astrobiology is study of earth". Astrobio.net. Retrieved 2010-10-15.

- "Introduction to Atmospheric Chemistry, by Daniel J. Jacob, Princeton University Press, 1999. Chapter 7, "The Greenhouse Effect"". Acmg.seas.harvard.edu. Retrieved 2010-10-15.

4

Methods of Controlling Greenhouse Gas Effects

The most popular methods of reducing greenhouse gas emissions are through policy-making that targets corporations and industries that have the largest carbon footprint. Placing a cap on carbon expenditure is an effective way to control greenhouse gases. Carbon credit, carbon offset and carbon neutrality are some of the topics in this chapter. Green energy is best understood in confluence with the major topics listed in the following chapter.

Carbon Credit

A carbon credit is a generic term for any tradable certificate or permit representing the right to emit one tonne of carbon dioxide or the mass of another greenhouse gas with a carbon dioxide equivalent (tCO_2e) equivalent to one tonne of carbon dioxide.

Carbon credits and carbon markets are a component of national and international attempts to mitigate the growth in concentrations of greenhouse gases (GHGs). One carbon credit is equal to one tonne of carbon dioxide, or in some markets, carbon dioxide equivalent gases. Carbon trading is an application of an emissions trading approach. Greenhouse gas emissions are capped and then markets are used to allocate the emissions among the group of regulated sources.

The goal is to allow market mechanisms to drive industrial and commercial processes in the direction of low emissions or less carbon intensive approaches than those used when there is no cost to emitting carbon dioxide and other GHGs into the atmosphere. Since GHG mitigation projects generate credits, this approach can be used to finance carbon reduction schemes between trading partners and around the world.

There are also many companies that sell carbon credits to commercial and individual customers who are interested in lowering their carbon footprint on a voluntary basis. These carbon offsetters purchase the credits from an investment fund or a carbon development company that has aggregated the credits from individual projects. Buyers and sellers can also use an exchange platform to trade, which is like a stock exchange for carbon credits. The quality of the credits is based in part on the validation process and sophistication of the fund or development company that acted as the sponsor to the carbon project. This is reflected in their price; voluntary units typically have less value than the units sold through the rigorously validated Clean Development Mechanism.

Definitions

The Collins English Dictionary defines a carbon credit as *"a certificate showing that a government* or company has paid to have a certain amount of carbon dioxide removed from the environment". The Environment Protection Authority of Victoria defines a carbon credit as a "generic term to assign a value to a reduction or offset of greenhouse gas emissions.. usually equivalent to one tonne of carbon dioxide equivalent (CO2-e)."

The Investopedia Inc investment dictionary defines a carbon credit as a "permit that allows the holder to emit one ton of carbon dioxide"..which "can be traded in the international market at their current market price".

Types

There are two main markets for carbon credits; Compliance Market credits Secondary / Verified Market credits (VERs)

Background

The burning of fossil fuels is a major source of greenhouse gas emissions, especially for power, cement, steel, textile, fertilizer and many other industries which rely on fossil fuels (coal, electricity derived from coal, natural gas and oil). The major greenhouse gases emitted by these industries are carbon dioxide, methane, nitrous oxide, hydrofluorocarbons (HFCs), etc., all of which increase the atmosphere's ability to trap infrared energy and thus affect the climate.

The concept of carbon credits came into existence as a result of increasing awareness of the need for controlling emissions. The IPCC (Intergovernmental Panel on Climate Change) has observed that:

Policies that provide a real or implicit price of carbon could create incentives for producers and consumers to significantly invest in low-GHG products, technologies and processes. Such policies could include economic instruments, government funding and regulation,

while noting that a tradable permit system is one of the policy instruments that has been shown to be environmentally effective in the industrial sector, as long as there are reasonable levels of predictability over the initial allocation mechanism and long-term price.

The mechanism was formalized in the Kyoto Protocol, an international agreement between more than 170 countries, and the market mechanisms were agreed through the subsequent Marrakesh Accords. The mechanism adopted was similar to the successful US Acid Rain Program to reduce some industrial pollutants.

Emission Allowances

Under the Kyoto Protocol, the 'caps' or quotas for Greenhouse gases for the developed Annex 1 countries are known as Assigned Amounts and are listed in Annex B. The quantity of the initial assigned amount is denominated in individual units, called Assigned amount units (AAUs), each of which represents an allowance to emit one metric tonne of carbon dioxide equivalent, and these are entered into the country's national registry.

In turn, these countries set quotas on the emissions of installations run by local business and other organizations, generically termed 'operators'. Countries manage this through their national registries, which are required to be validated and monitored for compliance by the UNFCCC. Each operator has an allowance of credits, where each unit gives the owner the right to emit one metric tonne of carbon dioxide or other equivalent greenhouse gas. Operators that have not used up their quotas can sell their unused allowances as carbon credits, while businesses that are about to exceed their quotas can buy the extra allowances as credits, privately or on the open market. As demand for energy grows over time, the total emissions must still stay within the cap, but it allows industry some flexibility and predictability in its planning to accommodate this.

By permitting allowances to be bought and sold, an operator can seek out the most cost-effective way of reducing its emissions, either by investing in 'cleaner' machinery and practices or by purchasing emissions from another operator who already has excess 'capacity'.

Since 2005, the Kyoto mechanism has been adopted for CO_2 trading by all the countries within the European Union under its European Trading Scheme (EU ETS) with the European Commission as its validating authority. From 2008, EU participants must link with the other developed countries who ratified Annex I of the protocol, and trade the six most significant anthropogenic greenhouse gases. In the United States, which has not ratified Kyoto, and Australia, whose ratification came into force in March 2008, similar schemes are being considered.

Kyoto's 'Flexible Mechanisms'

A tradable credit can be an emissions allowance or an assigned amount unit which was originally allocated or auctioned by the national administrators of a Kyoto-compliant cap-and-trade scheme, or it can be an offset of emissions. Such offsetting and mitigating activities can occur in any developing country which has ratified the Kyoto Protocol, and has a national agreement in place to validate its carbon project through one of the UNFCCC's approved mechanisms. Once approved, these units are termed Certified Emission Reductions, or CERs. The Protocol allows these projects to be constructed and credited in advance of the Kyoto trading period.

The Kyoto Protocol provides for three mechanisms that enable countries or operators in developed countries to acquire greenhouse gas reduction credits

- Under Joint Implementation (JI) a developed country with relatively high costs of domestic greenhouse reduction would set up a project in another developed country.

- Under the Clean Development Mechanism (CDM) a developed country can 'sponsor' a greenhouse gas reduction project in a developing country where the cost of greenhouse gas reduction project activities is usually much lower, but the atmospheric effect is globally equivalent. The developed country would be given credits for meeting its emission reduction targets, while the developing country would receive the capital investment and clean technology or beneficial change in land use.

- Under International Emissions Trading (IET) countries can trade in the international carbon credit market to cover their shortfall in Assigned amount units. Countries with surplus units can sell them to countries that are exceeding their emission targets under Annex B of the Kyoto Protocol.

These carbon projects can be created by a national government or by an operator within the country. In reality, most of the transactions are not performed by national governments directly, but by operators who have been set quotas by their country.

Emission Markets

For trading purposes, one allowance or CER is considered equivalent to one metric ton of CO_2 emissions. These allowances can be sold privately or in the international market at the prevailing market price. These trade and settle internationally and hence allow allowances to be transferred between countries. Each international transfer is validated by the UNFCCC. Each transfer of ownership within the European Union is additionally validated by the European Commission.

Climate exchanges have been established to provide a spot market in allowances, as well as futures and options market to help discover a market price and maintain liquidity. Carbon prices are normally quoted in Euros per tonne of carbon dioxide or its equivalent (CO_2e). Other greenhouse gasses can also be traded, but are quoted as standard multiples of carbon dioxide with respect to their global warming potential. These features reduce the quota's financial impact on business, while ensuring that the quotas are met at a national and international level.

Currently there are five exchanges trading in carbon allowances: the European Climate Exchange, NASDAQ OMX Commodities Europe, PowerNext, Commodity Exchange Bratislava and the European Energy Exchange. NASDAQ OMX Commodities Europe listed a contract to trade offsets generated by a CDM carbon project called Certified Emission Reductions (CERs). Many companies now engage in emissions abatement, offsetting, and sequestration programs to generate credits that can be sold on one of the exchanges. At least one private electronic market has been established in 2008: CantorCO2e. Carbon credits at Commodity Exchange Bratislava are traded at special platform - Carbon place.

Managing emissions is one of the fastest-growing segments in financial services in the City of London with a market estimated to be worth about €30 billion in 2007. Louis Redshaw, head of environmental markets at Barclays Capital predicts that "Carbon will be the world's biggest commodity market, and it could become the world's biggest market overall."

Setting a Market Price for Carbon

Unchecked, energy use and hence emission levels are predicted to keep rising over time. Thus the number of companies needing to buy credits will increase, and the rules of supply and demand will push up the market price, encouraging more groups to undertake environmentally friendly activities that create carbon credits to sell.

An individual allowance, such as an Assigned amount unit (AAU) or its near-equivalent European Union Allowance (EUA), may have a different market value to an offset such as a CER. This is due to the lack of a developed secondary market for CERs, a lack of homogeneity between projects which causes difficulty in pricing, as well as questions due to the principle of supplementarity and its lifetime. Additionally, offsets generated by a carbon project under the Clean Development Mechanism are potentially limited in value because operators in the EU ETS are restricted as to what percentage of their allowance can be met through these flexible mechanisms.

Yale University economics professor William Nordhaus argues that the price of carbon needs to be high enough to motivate the changes in behavior and changes in economic production systems necessary to effectively limit emissions of greenhouse gases.

Raising the price of carbon will achieve four goals. First, it will provide signals to consumers about what goods and services are high-carbon ones and should therefore be used more sparingly. Second, it will provide signals to producers about which inputs use more carbon (such as coal and oil) and which use less or none (such as natural gas or nuclear power), thereby inducing firms to substitute low-carbon inputs. Third, it will give market incentives for inventors and innovators to develop and introduce low-carbon products and processes that can replace the current generation of technologies. Fourth, and most important, a high carbon price will economize on the information that is required to do all three of these tasks. Through the market mechanism, a high carbon price will raise the price of products according to their carbon content. Ethical consumers today, hoping to minimize their "carbon footprint," have little chance of making an accurate calculation of the relative carbon use in, say, driving 250 miles as compared with flying 250 miles. A harmonized carbon tax would raise the price of a good proportionately to exactly the amount of CO_2 that is emitted in all the stages of production that are involved in producing that good. If 0.01 of a ton of carbon emissions results from the wheat growing and the milling and the trucking and the baking of a loaf of bread, then a tax of $30 per ton carbon will raise the price of bread by $0.30. The "carbon footprint" is automatically calculated by the price system. Consumers would still not know how much of the price is due to carbon emissions, but they could make their decisions confident that they are paying for the social cost of their carbon footprint.

Nordhaus has suggested, based on the social cost of carbon emissions, that an optimal price of carbon is around $30(US) per ton and will need to increase with inflation.

The social cost of carbon is the additional damage caused by an additional ton of carbon emissions. ... The optimal carbon price, or optimal carbon tax, is the market price (or carbon tax) on carbon emissions that balances the incremental costs of reducing carbon emissions with the incremental benefits of reducing climate damages. ... If a country wished to impose a carbon tax of $30 per ton of carbon, this would involve a tax on gasoline of about 9 cents per gallon. Similarly, the tax on coal-generated electricity would be about 1 cent per kWh, or 10 percent of the current retail price. At current levels of carbon emissions in the United States, a tax of $30 per ton of carbon would generate $50 billion of revenue per year.

How Buying Carbon Credits Can Reduce Emissions

Carbon credits create a market for reducing greenhouse emissions by giving a monetary value to the cost of polluting the air. Emissions become an internal cost of doing business and are visible on the balance sheet alongside raw materials and other liabilities or assets.

For example, consider a business that owns a factory putting out 100,000 tonnes of greenhouse gas emissions in a year. Its government is an Annex I country that enacts a law to limit the emissions that the business can produce. So the factory is given a quota of say 80,000 tonnes per year. The factory either reduces its emissions to 80,000 tonnes or is required to purchase carbon credits to offset the excess. After costing up alternatives the business may decide that it is uneconomical or infeasible to invest in new machinery for that year. Instead it may choose to buy carbon credits on the open market from organizations that have been approved as being able to sell legitimate carbon credits.

We should consider the impact of manufacturing alternative energy sources. For example, the energy consumed and the Carbon emitted in the manufacture and transportation of a large wind turbine would prohibit a credit being issued for a predetermined period of time.

- One seller might be a company that will offer to offset emissions through a project in the developing world, such as recovering methane from a swine farm to feed a power station that previously would use fossil fuel. So although the factory continues to emit gases, it would pay another group to reduce the equivalent of 20,000 tonnes of carbon dioxide emissions from the atmosphere for that year.

- Another seller may have already invested in new low-emission machinery and have a surplus of allowances as a result. The factory could make up for its emissions by buying 20,000 tonnes of allowances from them. The cost of the seller's new machinery would be subsidized by the sale of allowances. Both the buyer and the seller would submit accounts for their emissions to prove that their allowances were met correctly.

Credits Versus Taxes

Carbon credits and carbon taxes each have their advantages and disadvantages. Credits were chosen by the signatories to the Kyoto Protocol as an alternative to Carbon taxes. A criticism of tax-raising schemes is that they are frequently not hypothecated, and so some or all of the taxation raised by a government would be applied based on what the particular nation's government deems most fitting. However, some would argue that carbon trading is based around creating a lucrative artificial market, and, handled by free market enterprises as it is, carbon trading is not necessarily a focused or easily regulated solution.

By treating emissions as a market commodity some proponents insist it becomes easier for businesses to understand and manage their activities, while economists and traders can attempt to predict future pricing using market theories. Thus the main advantages of a tradeable carbon credit over a carbon tax are argued to be:

- the price may be more likely to be perceived as fair by those paying it. Investors in credits may have more control over their own costs.

- the flexible mechanisms of the Kyoto Protocol help to ensure that all investment goes into genuine sustainable carbon reduction schemes through an internationally agreed validation process.

- some proponents state that if correctly implemented a target level of emission reductions may somehow be achieved with more certainty, while under a tax the actual emissions might vary over time.

- it may provide a framework for rewarding people or companies who plant trees or otherwise meet standards exclusively recognized as "green."

The advantages of a carbon tax are argued to be:

- possibly less complex, expensive, and time-consuming to implement. This advantage is especially great when applied to markets like gasoline or home heating oil.

- perhaps some reduced risk of certain types of cheating, though under both credits and taxes, emissions must be verified.

- reduced incentives for companies to delay efficiency improvements prior to the establishment of the baseline if credits are distributed in proportion to past emissions.

- when credits are grandfathered, this puts new or growing companies at a disadvantage relative to more established companies.

- allows for more centralized handling of acquired gains

- worth of carbon is stabilized by government regulation rather than market fluctuations. Poor market conditions and weak investor interest have a lessened impact on taxation as opposed to carbon trading.

Creating Carbon Credits

The principle of Supplementarity within the Kyoto Protocol means that internal abatement of emissions should take precedence before a country buys in carbon credits. However it also established the Clean Development Mechanism as a Flexible Mechanism by which capped entities could develop measurable and permanent emissions reductions voluntarily in sectors outside the cap. Many criticisms of carbon credits stem from the fact that establishing that an emission of CO_2-equivalent greenhouse gas has truly been reduced involves a complex process. This process has evolved as the concept of a carbon project has been refined over the past 10 years.

The first step in determining whether or not a carbon project has legitimately led to the reduction of measurable and permanent emissions is understanding the CDM methodology process. This is the process by which project sponsors submit, through a Designated Operational Entity (DOE), their concepts for emissions reduction creation. The CDM Executive Board, with the CDM Methodology Panel and their expert advisors, review each project and decide how and if they do indeed result in reductions that are additional

Additionality and its Importance

It is also important for any carbon credit (offset) to prove a concept called additionality. The concept of additionality addresses the question of whether the project would have happened in the absence of an intervention in the form of the price signal of carbon credits. Only projects with emissions below their baseline level, defined as emissions under a scenario without this price signal (holding all other factors constant), represent a net environmental benefit. Carbon projects that yield strong financial returns even in the absence of revenue from carbon credits; or that are compelled by regulations; or that represent common practice in an industry; are usually not considered additional. A full determination of additionality requires a careful investigation of proposed carbon offset projects.

It is generally agreed that voluntary carbon offset projects must demonstrate additionality to ensure the legitimacy of the environmental stewardship claims resulting from the retirement of carbon credits (offsets).

Criticisms

The Kyoto mechanism is the only internationally agreed mechanism for regulating carbon credit activities, and, crucially, includes checks for additionality and overall effectiveness. Its supporting organisation, the UNFCCC, is the only organisation with a global mandate on the overall effectiveness of emission control systems, although enforcement of decisions relies on national co-operation. The Kyoto trading period only applies for five years between 2008 and 2012. The first phase of the EU ETS system started before then, and is expected to continue in a third phase afterwards, and may co-ordinate with whatever is internationally agreed at but there is general uncertainty as to what will be agreed in Post–Kyoto Protocol negotiations on greenhouse gas emissions. As business investment often operates over decades, this adds risk and uncertainty to their plans. As several countries responsible for a large proportion of global emissions (notably USA, India, China) have avoided mandatory caps, this also means that businesses in capped countries may perceive themselves to be working at a competitive disadvantage against those in uncapped countries as they are now paying for their carbon costs directly.

A key concept behind the cap and trade system is that national quotas should be chosen to represent genuine and meaningful reductions in national output of emissions. Not only does this ensure that overall emissions are reduced but also that the costs of emissions trading are carried fairly across all parties to the trading system. However, governments of capped countries may seek to unilaterally weaken their commitments, as evidenced by the 2006 and 2007 National Allocation Plans for several countries in the EU ETS, which were submitted late and then were initially rejected by the European Commission for being too lax.

A question has been raised over the grandfathering of allowances. Countries within the EU ETS have granted their incumbent businesses most or all of their allowances for free. This can sometimes be perceived as a protectionist obstacle to new entrants into their markets. There have also been accusations of power generators getting a 'windfall' profit by passing on these emissions 'charges' to their customers. As the EU ETS moves into its second phase and joins up with Kyoto, it seems likely that these problems will be reduced as more allowances will be auctioned.

Carbon Neutrality

Carbon neutrality, or having a net zero carbon footprint, refers to achieving net zero carbon emissions by balancing a measured amount of carbon released with an equivalent amount sequestered or offset, or buying enough carbon credits to make up the difference. It is used in the context of carbon dioxide releasing processes associated with transportation, energy production, and industrial processes such as production of carbon neutral fuel.

The carbon neutrality concept may be extended to include other greenhouse gases (GHG) measured in terms of their carbon dioxide equivalence (CO_2e) —the impact a GHG has on the atmosphere expressed in the equivalent amount of CO_2. The term "climate neutral" reflects the broader inclusiveness of other greenhouse gases in climate change, even if CO_2 is the most abundant, encompassing other greenhouse gases regulated by the Kyoto Protocol, namely: methane (CH_4), nitrous oxide (N_2O), hydrofluorocarbons (HFC), perfluorocarbons (PFC), and sulphur hexafluoride (SF_6). Both terms are used interchangeably throughout this article.

The best practice for organizations and individuals seeking carbon neutral status entails reducing and/or avoiding carbon emissions first so that only unavoidable emissions are offset. Carbon neutral status is commonly achieved in two ways:

- Balancing carbon dioxide released into the atmosphere from burning fossil fuels, with renewable energy that creates a similar amount of useful energy, so that the carbon emissions are compensated, or alternatively using only renewable energies that don't produce any carbon dioxide (also called a post-carbon economy).

- Carbon offsetting by paying others to remove or sequester 100% of the carbon dioxide emitted from the atmosphere – for example by planting trees – or by funding 'carbon projects' that should lead to the prevention of future greenhouse gas emissions, or by buying carbon credits to remove (or 'retire') them through carbon trading. While carbon offsetting is often used alongside energy conservation measures to minimize energy use, the practice is criticized by some.

The concept may be extended to include other greenhouse gases measured in terms of their carbon dioxide equivalence. The phrase was the *New Oxford American Dictionary's* Word Of The Year for 2006.

Process

Carbon, or climate, neutrality is usually achieved by combining the following steps (although these may vary depending whether the strategy is implemented by individuals, companies, organizations, cities, regions, or countries):

Commitment

In the case of individuals, decision-making is likely to be straightforward, but for more complex set-ups, it usually requires political leadership at the highest level and wide popular agreement that the effort is worth making.

Counting and Analyzing

Counting and analyzing the emissions that need to be eliminated, and the options for doing so, is the most crucial step in the cycle as it enables setting the priorities for action – from the products purchased to energy use and transport – and to start monitoring progress. This can be achieved through a GHG inventory that aims at answering questions such as:

- Which operations, activities, units should be included?

- Which sources should be included?

- Who is responsible for which emissions?

- Which gases should be included?

For individuals, carbon calculators simplify compiling an inventory. Typically they measure electricity consumption in kWh, the amount and type of fuel used to heat water and warm the house, and how many kilometres an individual drives, flies and rides in different vehicles. Individuals may also set various limits of the system they are concerned with, e.g. personal GHG emissions, household emissions, or the company they work for.There are plenty of carbon calculators available online, which vary significantly in their usefulness and the parameters they measure. Some, for example, factor in only cars, aircraft and household energy use. Others cover household waste or leisure interests as well. In some circumstances, actually going beyond carbon neutral (usually after a certain length of time taken to reach carbon breakeven) is an objective.

Action

In starting to work towards climate neutrality, businesses and local administrations can make use of an environmental (or sustainability) management system or EMS established by the international standard ISO 14001 (developed by the International Organization for Standardization). Another EMS framework is EMAS, the European Eco Management and Audit Scheme, used by numerous companies throughout the EU. Many local authorities apply the management system to certain sectors of their administration or certify their whole operations.

Reduction

One of the strongest arguments for reducing GHG emissions is that it will often save money. Energy prices across the world are rising, making it harder to afford to travel, heat and light homes and factories, and keep a modern economy ticking over. So it is both common sense and sensible for the climate to use energy as sparingly as possible. Examples of possible actions to reduce GHG emissions are:

- Limiting energy usage and emissions from transportation (walking, using bicycles or public transport, avoiding flying, using low-energy vehicles), as well as from buildings, equipment, animals and processes.

- Obtaining electricity and other energy from a renewable energy source, either directly by generating it (installing solar panels on the roof for example) or by selecting an approved green energy provider, and by using low-carbon alternative fuels such as sustainable biofuels.

Offsetting

The use of Carbon offsets aims to neutralize a certain volume of GHG emissions by funding projects which should cause an equivalent reduction of GHG emissions somewhere else, such as tree planting. Under the premise "First reduce what you can, then offset the remainder", offsetting can be done by supporting a responsible carbon project, or by buying carbon offsets or carbon credits.

Carbon offsetting is also a tool for severals local authorities in the world.

Offsetting is sometimes seen as a charged and contentious issue. For example, James Hansen describes offsets as "modern day indulgences, sold to an increasingly carbon-conscious public to absolve their climate sins."

Evaluation and Repeating

This phase includes evaluation of the results and compilation of a list of suggested improvements, with results documented and reported, so that experience gained of what does (and does not) work is shared with those who can put it to good use.Finally, with all that completed, the cycle starts all over again, only this time incorporating the lessons learnt. Science and technology move on, regulations become tighter, the standards people demand go up. So the second cycle will go further than the first, and the process will continue, each successive phase building on and improving on what went before.

Being carbon neutral is increasingly seen as good corporate or state social responsibility and a growing list of corporations and states are announcing dates for when they intend to become fully neutral. Events such as the G8 Summit and organizations like the World Bank are also using offset schemes to become carbon neutral. Artists like The Rolling Stones and Pink Floyd have made albums or tours carbon neutral.

Direct and Indirect Emissions

To be considered carbon neutral, an organization must reduce its carbon footprint to zero. Determining what to include in the carbon footprint depends upon the organization and the standards they are following.

Generally, direct emissions sources must be reduced and offset completely, while indirect emissions from purchased electricity can be reduced with renewable energy purchases.

Direct emissions include all pollution from manufacturing, company owned vehicles and reimbursed travel, livestock and any other source that is directly controlled by the owner. Indirect emissions include all emissions that result from the use or purchase of a product. For instance, the direct emissions of an airline are all the jet fuel that is burned, while the indirect emissions include manufacture and disposal of airplanes, all the electricity used to operate the airline's office, and the daily emissions from employee travel to and from work. In another example, the power company has a direct emission of greenhouse gas, while the office that purchases it considers it an indirect emission.

Simplification of Standards and Definitions

Carbon neutral fuels are those that neither contribute to nor reduce the amount of carbon into the atmosphere. Before an agency can certify an organization or individual as carbon neutral, it is important to specify whether indirect emissions are included in the Carbon Footprint calculation. Most Voluntary Carbon neutral certifiers such as Standard Carbon in the US, require both direct and indirect sources to be reduced and offset. As an example, for an organization to be certified carbon neutral by Standard Carbon, it must offset all direct and indirect emissions from travel by 1 lb CO_2e per passenger mile, and all non-electricity direct emissions 100%. Indirect electrical purchases must be equalized either with offsets, or renewable energy purchase. This standard differs slightly from the widely used World Resource Institute and may be easier to calculate and apply.

Much of the confusion in carbon neutral standards can be attributed to the number of voluntary carbon standards which are available. For organizations looking at which carbon offsets to purchase, knowing which standards are robust, credible in permanent is vital in choosing the right carbon offsets and projects to get involved in. Some of the main standards in the voluntary market include; CEB VER Standard, The Voluntary Carbon Standard, The Gold Standard and The California Climate Action Registry. In addition companies can purchase Certified Emission Reductions (CERs) which result from mitigated carbon emissions from UNFCCC approved projects for voluntary purposes. The concept of shared resources also reduces the volume of carbon a particular organization has to offset, with all upstream and downstream emissions the responsibility of other organizations or individuals. If all organizations and individuals were involved then this would not result in any double accounting.

Regarding terminology in UK and Ireland, in December 2011 the Advertising Standards Authority (in an ASA decision which was upheld by its Independent Reviewer, Sir Hayden Phillips) controversially ruled that no manufactured product can be marketed as "zero carbon", because carbon was inevitably emitted during its manufacture. This decision was made in relation to a solar panel system whose embodied carbon was repaid during 1.2 years of use and it appears to mean that no buildings or manufactured products can legitimately be described as zero carbon in its jurisdiction.

Pledges

Being carbon neutral is increasingly seen as good corporate or state social responsibility and a growing list of corporations, cities and states are announcing dates for when they intend to become fully neutral.

Companies and Organizations

The original Climate Neutral Network was an Oregon-based non-profit organization founded by Sue Hall and incorporated in 1999 to persuade companies that being climate neutral was potentially cost saving as well as environmentally sustainable. It developed both the Climate Neutral Certification and Climate Cool brand name with key stakeholders such as the US EPA, The Nature Conservancy, Rocky Mountain Institute, Conservation International, and the World Resources Institute and succeeded in enrolling the 2002 Winter Olympics to compensate for its associated greenhouse gas emissions. The non-profit's web site as of March 2011, lists the organization as closing its doors and plans to continue the Climate Cool upon transfer to a new non-profit organization, unknown at this time. Interestingly, the for-profit consulting firm, Climate Neutral Business Network, lists the same Sue Hall as CEO and lists many of the same companies who were participants in the original Climate Neutral Network, as consulting clients.

Few companies have actually attained Climate Neutral Certification, applying to a rigorous review process and establishing that they have achieved absolute net zero or better impact on the world's climate. Shaklee Corporation announced it became the first Climate Neutral certified company in April 2000. Climate Neutral Business Network states that it certified Dave Matthews Band's concert tour as Climate Neutral. The Christian Science Monitor criticized the use of NativeEnergy. a for-profit company that sells offsets credits to businesses and celebrities like Dave Matthews.

Salt Spring Coffee has become carbon neutral by lowering emissions through reducing long-range trucking and using bio-diesel fuel in delivery trucks, upgrading to energy efficient equipment and purchasing carbon offsets. The company claims to the first carbon neutral coffee sold in Canada. Salt Spring Coffee was recognized by the David Suzuki Foundation in their 2010 report *Doing Business in a New Climate*.

Some corporate examples of self-proclaimed carbon neutral and climate neutral initiatives include Dell, Google, HSBC, ING Group, PepsiCo, Sky, Tesco, Toronto-Dominion Bank, Asos and Bank of Montreal.

Under the leadership of Secretary-General Ban Ki-moon, the United Nations pledged to work towards climate neutrality in December 2007. The United Nations Environment Programme (UNEP) announced it was becoming climate neutral in 2008 and established a Climate Neutral Network to promote the idea in February 2008.

Events such as the G8 Summit and organizations like the World Bank are also using offset schemes to become carbon neutral. Artists like The Rolling Stones and Pink Floyd have made albums or tours carbon neutral, while Live Earth says that its seven concerts held on 7 July 2007 were the largest carbon neutral public event in history.

The Vancouver 2010 Olympic and Paralympic Winter Games were the first carbon neutral Games in history.

Buildings are the largest single contributor to the production of greenhouse gases. The American Institute of Architects 2030 Commitment is a voluntary program for AIA member firms and other entities in the built environment that asks these organizations to pledge to design all their buildings to be carbon neutral by 2030.

In 2010, architectural firm HOK worked with energy and daylighting consultant The Weidt Group to design a 170,735-square-foot (15,861.8 m²) net zero carbon emissions Class A office building prototype in St. Louis, Missouri, U.S.

Costa Rica

Costa Rica aims to be fully carbon neutral by 2021. In 2004, 46.7% of Costa Rica's primary energy came from renewable sources, while 94% of its electricity was generated from hydroelectric power, wind farms and geothermal energy in 2006. A 3.5% tax on gasoline in the country is used for payments to compensate landowners for growing trees and protecting forests and its government is making further plans for reducing emissions from transport, farming and industry.

Denmark

Samsø island in Denmark is the largest carbon-neutral settlement on the planet, with a population of 4200, based on wind-generated electricity and biomass-based district heating. They currently generate extra wind power and export the electricity to compensate for petro-fueled vehicles.

There are future hopes of using electric or biofuel vehicles.

Maldives

The ex-president of the Maldives has pledged to make his country carbon-neutral within a decade by moving to wind and solar energy. The Maldives, a country consisting of very low-lying islands, would be one of the first countries to be submerged due to sea level rise. The Maldives presided over the foundation of the Climate Vulnerable Forum.

New Zealand

Another nation to pledge carbon neutrality is New Zealand. Its Carbon Neutral Public Sector Initiative aimed to offset the greenhouse gas emissions of an initial group of six governmental agencies by 2012. Unavoidable emissions would be offset, primarily through indigenous forest regeneration projects on conservation land. All 34 public service agencies also needed to have emission reduction plans in place. The Carbon Neutral Public Service programme was discontinued in March 2009.

Norway

On April 19, 2007, Prime Minister Jens Stoltenberg announced to the Labour Party annual congress that Norway's greenhouse gas emissions would be cut by 10 percent more than its Kyoto commitment by 2012, and that the government had agreed to achieve emission cuts of 30% by 2020. He also proposed that Norway should become carbon neutral by 2050, and called upon other rich countries to do likewise. This carbon neutrality would be achieved partly by carbon offsetting, a proposal criticised by Greenpeace, who also called on Norway to take responsibility for the 500m tonnes of emissions caused by its exports of oil and gas. World Wildlife Fund Norway also believes that the purchase of carbon offsets is unacceptable, saying 'it is a political stillbirth to believe that China will quietly accept that Norway will buy climate quotas abroad'. The Norwegian environmental activist Bellona Foundation believes that the prime minister was forced to act due to pressure from anti-European Union members of the coalition government, and called the announcement 'visions without content'. In January 2008 the Norwegian government went a step further and declared a goal of being carbon neutral by 2030. But the government has not been specific about any plans to reduce emissions at home; the plan is based on buying carbon offsets from other countries, and very little has actually been done to reduce Norway's emissions, apart from a very successful policy for electric vehicles

Spain

In Spain, in 2014, the island of El Hierro became carbon neutral (for its power production). Also, the city of Logroño Montecorvo in La Rioja will be carbon neutral once completed.

Iceland

Iceland is also moving towards climate neutrality. Over 99% of electricity production and almost 80% of total energy production comes from hydropower and geothermal. No other nation uses such a high proportion of renewable energy resources. In February 2008, Costa Rica, Iceland, New Zealand and Norway were the first four countries to join the Climate Neutral Network, an initiative led by the United Nations Environment Programme (UNEP) to catalyze global action towards low carbon economies and societies.

Vatican City

In July 2007, Vatican City announced a plan to become the first carbon-neutral state in the world, following the politics of the Pope to eliminate global warming. The goal would be reached through the donation of the Vatican Climate Forest in Hungary. The forest is to be sized to offset the year's carbon dioxide emissions. However, no trees have actually been planted as of 2008. The company KlimaFa is no longer in existence and hasn't fulfilled its promises. In November 2008, the city state also installed and put into operation 2,400 solar panels on the roof of the Paul VI Centre audience hall. In 2013, Vatican City became the world's first carbon neutral country

British Columbia

In June 2011, the Canadian Province of British Columbia announced they had officially become the first provincial/state jurisdiction in North America to achieve carbon neutrality in public sector operations: every school, hospital, university, Crown corporation, and government office measured, reported and purchased carbon offsets on all their 2010 Greenhouse Gas emissions as required under legislation. Local Governments across B.C. are also beginning to declare Carbon Neutrality.

Sweden

2050, Sweden will no longer contribute to the greenhouse effect. The vision is that net greenhouse gas emissions should be zero. The overall objective is that the increase in global temperature should be limited to two degrees, and that the concentration of greenhouse gases in the atmosphere stabilizes at a maximum of 400 ppm.

Carbon Neutral Initiatives

Many initiatives seek to assist individuals, businesses and states in reducing their carbon footprint or achieving climate neutrality. These include website neutralization projects like CO_2Stats and, the similar European initiative CO_2 neutral website as well as the Climate Neutral Network, Caring for Climate, and Together campaign.

Certification

Although there is currently no international certification scheme for carbon or climate neutrality, some countries have established national certification schemes. Examples include Norwegian Eco-Lighthouse Program and the Australian government's National Carbon Offset Standard (NCOS).

Certifications are also available from the CEB, BSI (PAS 2060) and The CarbonNeutral Company (CarbonNeutral).

Climate Neutral Certification

Climate Neutral Certification was established and trademarked originally through the Climate Neutral Network, an Oregon-based non-profit organization, not to be confused with the UNEP's Climate Neutral Network. Applications for certification are no longer being accepted according to the non-profit organization's web site, where the organization also states it is closing its doors. The first three companies certified as Climate Neutral were Shaklee Corporation, Interface, Inc., and Saunders Hotels. Stakeholders in developing and supporting the Climate Neutral Certification are listed as The Nature Conservancy, Conservation International, Rocky Mountain Institute, and the U.S EPA. What is unclear is whether the Climate Neutral certification will be continued by the for-profit consulting firm, Climate Neutral Business Network, or another non-profit organization in the future.

Climate Neutral Network also promoted, trademarked, and licensed the brand Climate Cool, for products certified by the organization's Environmental Review Panel and determined to achieve net zero climate impact, by reducing and offsetting associated emissions. The organization's web site promises to transfer the Climate Cool branding to another non-profit organization, upon closing the current organization.

In Australia, the government-endorsed carbon neutral certification is the National Carbon Offset Standard (NCOS).

Carbon Offset

A carbon offset is a reduction in emissions of carbon dioxide or greenhouse gases made in order to compensate for or to offset an emission made elsewhere.

Carbon offsets are measured in metric tons of carbon dioxide-equivalent (CO_2e) and may represent six primary categories of greenhouse gases: carbon dioxide (CO_2), methane (CH_4), nitrous oxide (N_2O), perfluorocarbons (PFCs), hydrofluorocarbons (HFCs), and sulfur hexafluoride (SF_6). One carbon offset represents the reduction of one metric ton of carbon dioxide or its equivalent in other greenhouse gases.

There are two markets for carbon offsets. In the larger, compliance market, companies, governments, or other entities buy carbon offsets in order to comply with caps on the total amount of carbon dioxide they are allowed to emit. This market exists in order to achieve compliance with obligations of Annex 1 Parties under the Kyoto Protocol, and of liable entities under the EU Emission Trading Scheme. In 2006, about \$5.5 billion of carbon offsets were purchased in the compliance market, representing about 1.6 billion metric tons of CO_2e reductions.

In the much smaller, voluntary market, individuals, companies, or governments purchase carbon offsets to mitigate their own greenhouse gas emissions from transportation, electricity use, and other sources. For example, an individual might purchase carbon offsets to compensate for the greenhouse gas emissions caused by personal air travel. Many companies offer carbon offsets as an up-sell during the sales process so that customers can mitigate the emissions related with their product or service purchase (such as offsetting emissions related to a vacation flight, car rental, hotel stay, consumer good, etc.). In 2008, about $705 million of carbon offsets were purchased in the voluntary market, representing about 123.4 million metric tons of CO_2e reductions. Some fuel suppliers in the UK offer fuel which has been carbon offset such as Fuel dyes.

Offsets are typically achieved through financial support of projects that reduce the emission of greenhouse gases in the short- or long-term. The most common project type is renewable energy, such as wind farms, biomass energy, or hydroelectric dams. Others include energy efficiency projects, the destruction of industrial pollutants or agricultural byproducts, destruction of landfill methane, and forestry projects. Some of the most popular carbon offset projects from a corporate perspective are energy efficiency and wind turbine projects.

Carbon offsetting has gained some appeal and momentum mainly among consumers in western countries who have become aware and concerned about the potentially negative environmental effects of energy-intensive lifestyles and economies. The Kyoto Protocol has sanctioned offsets as a way for governments and private companies to earn carbon credits that can be traded on a marketplace. The protocol established the Clean Development Mechanism (CDM), which validates and measures projects to ensure they produce authentic benefits and are genuinely "additional" activities that would not otherwise have been undertaken. Organizations that are unable to meet their emissions quota can offset their emissions by buying CDM-approved Certified Emissions Reductions. Emissions from burning fuel, such as red diesel, has pushed one UK fuel supplier to create a carbon offset fuel named Carbon Offset Red Diesel.

Offsets may be cheaper or more convenient alternatives to reducing one's own fossil-fuel consumption. However, some critics object to carbon offsets, and question the benefits of certain types of offsets. Due diligence is recommended to help businesses in the assessment and identification of "good quality" offsets to ensure offsetting provides the desired additional environmental benefits, and to avoid reputational risk associated with poor quality offsets.

Offsets are viewed as an important policy tool to maintain stable economies. One of the hidden dangers of climate change policy is unequal prices of carbon in the economy, which can cause economic collateral damage if production flows to regions or industries that have a lower price of carbon—unless carbon can be purchased from that area, which offsets effectively permit, equalizing the price.

Wind turbines near Aalborg, Denmark. Renewable energy projects are the most common source of carbon offsets.

Definitions

The World Resources Institute defines a carbon offset as "a unit of carbon dioxide-equivalent (CO_2e) that is reduced, avoided, or sequestered to compensate for emissions occurring elsewhere".

The Collins English Dictionary defines a carbon offset as "a compensatory measure made by an individual or company for carbon emissions, usually through sponsoring activities or projects which increase carbon dioxide absorption, such as tree planting".

The Environment Protection Authority of Victoria (Australia) defines a carbon offset as: "a monetary investment in a project or activity elsewhere that abates greenhouse gas (GHG) emissions or sequesters carbon from the atmosphere that is used to compensate for GHG emissions from your own activities. Offsets can be bought by a business or individual in the voluntary market (or within a trading scheme), a carbon offset usually represents one tonne of CO_2-e".

The Stockholm Environment Institute defines a carbon offset as "a credit for negating or diminishing the impact of emitting a ton of carbon dioxide by paying someone else to absorb or avoid the release of a ton of CO_2 elsewhere".

The University of Oxford Environmental Change Institute defines a carbon offset as "mechanism whereby individuals and corporations pay for reductions elsewhere in order to offset their own emissions".

The Encyclopædia Britannica defines a carbon offset as *"any activity that compensates for the emission of carbon dioxide (CO_2) or other greenhouse gases (measured in carbon dioxide equivalents [CO_2e]) by providing for an emission reduction elsewhere."*

A carbon offset is a greenhouse gas (GHG) reduction that is used to counterbalance, or offset, a GHG emission. They are sometimes also referred to as carbon credits, VERs, or CERs.

A carbon offset occurs when an individual or organization emits a given amount of GHG emissions but invests in measures that remove the equivalent volume of GHG emissions from the atmosphere or prevent the emissions from taking place at all. Carbon offsets are a financial instrument that represents this reduction in GHG emissions.

Features

Carbon offsets have several common features:

- *Vintage*. The vintage is the year in which the carbon reduction takes place.

- *Source*. The source refers to the project or technology used in offsetting the carbon emissions. Projects can include land-use, methane, biomass, renewable energy and industrial energy efficiency. Projects may also have secondary benefits (co-benefits). For example, projects that reduce agricultural greenhouse gas emissions may improve water quality by reducing fertilizer usage.

- *Certification regime*. The certification regime describes the systems and procedures that are used to certify and register carbon offsets. Different methodologies are used for measuring and verifying emissions reductions, depending on project type, size and location. For example, the CDM uses another. In the voluntary market, a variety of industry standards exist. These include the Voluntary Carbon Standard and the CDM Gold Standard that are implemented to provide third-party verification of carbon offset projects. There are some additional standards for the validation of co-benefits, including the CCBS, issued by the Climate, Community & Biodiversity Alliance and the Social Carbon Standard, issued by Ecologica Institute.

Carbon Offset Markets

Global Market

In 2009, 8.2 billion metric tons of carbon dioxide equivalent changed hands worldwide, up 68 per cent from 2008, according to the study by carbon-market research firm Point Carbon, of Washington and Oslo. But at EUR94 billion, or about $135 billion, the market's value was nearly unchanged compared with 2008, with world carbon prices averaging EUR11.40 a ton, down about 40 per cent from the previous year, according to the study. The World Bank's "State and Trends of the Carbon Market 2010" put the overall value of the market at $144 billion, but found that a significant part of this figure resulted from manipulation of a VAT loophole.

· 90% of voluntary offset volumes were contracted by the private sector – where corporate social responsibility and industry leadership were primary motivations for offset purchases.

· Offset buyers' desire to positively impact the climate resilience of their supply chain or sphere of influence was evident in our data which identifies a strong relationship between buyers' business sectors and the project categories from which they contract offsets.

E.U. Market

The global carbon market is dominated by the European Union, where companies that emit greenhouse gases are required to cut their emissions or buy pollution allowances or carbon credits from the market, under the European Union Emission Trading Scheme (EU ETS). Europe, which has seen volatile carbon prices due to fluctuations in energy prices and supply and demand, will continue to dominate the global carbon market for another few years, as the U.S. and China—the world's top polluters—have yet to establish mandatory emission-reduction policies.

U.S. Market

On the whole, the U.S. market remains primarily a voluntary market, but multiple cap and trade regimes are either fully implemented or near-imminent at the regional level. The first mandatory, market-based cap-and-trade program to cut CO_2 in the U.S., called the Regional Greenhouse Gas Initiative (RGGI), kicked into gear in Northeastern states in 2009, growing nearly tenfold to $2.5 billion, according to Point Carbon. Western Climate Initiative (WCI)—a regional cap-and-trade program including seven western states (California notably among them) and four Canadian provinces—has established a regional target for reducing heat-trapping emissions of 15 percent below 2005 levels by 2020. A component of California's own Global Warming Solutions Act of 2006, kicked off in early 2013, requires high-emissions industries to purchase carbon credits to cover emissions in excess of 25,000 CO_2 metric tons.

Voluntary Market

Participants

A wide range of participants are involved in the voluntary market, including providers of different types of offsets, developers of quality assurance mechanisms, third party verifiers, and consumers who purchase offsets from domestic or international providers. Suppliers include for-profit companies, governments, charitable non-governmental organizations, colleges and universities, and other groups.

Motivations

According to industry analyst Ecosystem Marketplace, the voluntary markets present the opportunity for citizen consumer action, as well as an alternative source of carbon finance and an incubator for carbon market innovation. In their survey of voluntary markets, data has shown that "Corporate Social Responsibility" and "Public Relations/Branding" are clearly in first place among motivations for voluntary offset purchases, with evidence indicating that companies seek to offset emissions "for goodwill, both of the general public and their investors".

In addition, regarding market composition, research indicates: "Though many analysts perceive pre-compliance buying as a dominant driving force in the voluntary market, the results of our survey have repeatedly indicated that precompliance motives (as indicated by 'investment/resale and 'anticipation of regulation') remain secondary to those of the pure voluntary market (companies/individuals offsetting their emissions)."

Pre-compliance & Trading

The other main category of buyers on the voluntary markets are those engaged in pre-compliance and/or trading. Those purchasing offsets for pre-compliance purposes are doing so with the expectation, or as a hedge against the possibility, of future mandatory cap and trade regulations. As a mandatory cap would sharply increase the price of offsets, firms—especially those with large carbon footprints and the corresponding financial exposure to regulation—make the decision to acquire offsets in advance at what are expected to be lower prices.

The trading market in offsets in general resembles the trade in other commodities markets, with financial professionals including hedge funds and desks at major investment banks, taking positions in the hopes of buying cheap and selling dear, with their motivation typically short or medium term financial gain.

Retail

Multiple players in the retail market have offerings that enable consumers and businesses to calculate their carbon footprint, most commonly through a web-based interface including a calculator or questionnaire, and sell them offsets in the amount of that footprint. In addition many companies selling products and services, especially carbon-intensive ones such as airline travel, offer options to bundle a proportional offsetting amount of carbon credits with each transaction.

Suppliers of voluntary offsets operate under both nonprofit and social enterprise models, or a blended approach sometimes referred to as triple bottom line. Other suppliers include broader environmentally focused organizations with website subsections or initiatives that enable retail voluntary offset purchases by members, and government created projects.

Features of Companies that Voluntarily Offset Emissions

Companies that voluntarily offset their own emissions tend to be of relatively low carbon intensity, as they can offset a significant proportion of their emissions at relatively low cost. Voluntary offsetting is particularly common in the financial sector. 61 per cent of financial companies in the FTSE 100 offset at least a portion of their 2009 emissions. Twenty-two per cent of financial companies in the FTSE 100 considered their entire 2009 operations to be carbon neutral.

Sources of Carbon Offsets

The CDM identifies over 200 types of projects suitable for generating carbon offsets, which are grouped into broad categories. These project types include renewable energy, methane abatement, energy efficiency, reforestation and fuel switching.

Renewable Energy

Renewable energy offsets commonly include wind power, solar power, hydroelectric power and biofuel. Some of these offsets are used to reduce the cost differential between renewable and conventional energy production, increasing the commercial viability of a choice to use renewable energy sources.

Renewable Energy Credits (RECs) are also sometimes treated as carbon offsets, although the concepts are distinct. Whereas a carbon offset represents a reduction in greenhouse gas emissions, a REC represents a quantity of energy produced from renewable sources. To convert RECs into offsets, the clean energy must be translated into carbon reductions, typically by assuming that the clean energy is displacing an equivalent amount of conventionally produced electricity from the local grid. This is known as an indirect offset (because the reduction doesn't take place at the project site itself, but rather at an external site), and some controversy surrounds the question of whether they truly lead to "additional" emission reductions and who should get credit for any reductions that may occur. Intel corporation is the largest purchaser of renewable power in the US.

Methane Collection and Combustion

Some offset projects consist of the combustion or containment of methane generated by farm animals (by use of an anaerobic digester), landfills or other industrial waste. Methane has a global warming potential (GWP) 23 times that of CO_2; when combusted, each molecule of methane is converted to one molecule of CO_2, thus reducing the global warming effect by 96%.

An example of a project using an anaerobic digester can be found in Chile where in December

2000, the largest pork production company in Chile, initiated a voluntary process to implement advanced waste management systems (anaerobic and aerobic digestion of hog manure), in order to reduce greenhouse gas (GHG) emissions.

Energy Efficiency

Chicago Climate Justice activists protesting cap and trade legislation in front of Chicago Climate Exchange building in Chicago Loop

While carbon offsets that fund renewable energy projects help lower the carbon intensity of energy *supply*, energy conservation projects seek to reduce the overall *demand* for energy. Carbon offsets in this category fund projects of several types:

1. Cogeneration plants generate both electricity and heat from the same power source, thus improving upon the energy efficiency of most power plants, which waste the energy generated as heat.

2. Fuel efficiency projects replace a combustion device with one using less fuel per unit of energy provided. Assuming energy demand does not change, this reduces the carbon dioxide emitted.

3. Energy-efficient buildings reduce the amount of energy wasted in buildings through efficient heating, cooling or lighting systems. In particular, the replacement of incandescent light bulbs with compact fluorescent lamps can have a drastic effect on energy consumption. New buildings can also be constructed using less carbon-intensive input materials.

Destruction of Industrial Pollutants

Industrial pollutants such as hydrofluorocarbons (HFCs) and perfluorocarbons (PFCs) have a GWP many thousands of times greater than carbon dioxide by volume. Because these pollutants are easily captured and destroyed at their source, they present a large and low-cost source of carbon offsets. As a category, HFCs, PFCs, and N_2O reductions represent 71 per cent of offsets issued under the CDM.

Land use, Land-use Change and Forestry

Land use, land-use change and forestry (LULUCF) projects focus on natural carbon sinks such as forests and soil. Deforestation, particularly in Brazil, Indonesia and parts of Africa, account for about 20 per cent of greenhouse gas emissions. Deforestation can be avoided either by paying directly for forest preservation, or by using offset funds to provide substitutes for forest-based products. There is a class of mechanisms referred to as REDD schemes (Reducing emissions from deforestation and forest degradation), which may be included in a post-Kyoto agreement. REDD credits provide carbon offsets for the protection of forests, and provide a possible mechanism to allow funding from developed nations to assist in the protection of native forests in developing nations.

Almost half of the world's people burn wood (or fiber or dung) for their cooking and heating needs. Fuel-efficient cook stoves can reduce fuel wood consumption by 30 to 50%, though the warming of the earth due to decreases in particulate matter (i.e. smoke) from such fuel-efficient stoves has not been addressed. There are a number of different types of LULUCF projects:

- Avoided deforestation is the protection of existing forests.

- Reforestation is the process of restoring forests on land that was once forested.

- Afforestation is the process of creating forests on land that was previously unforested, typically for longer than a generation.

- Soil management projects attempt to preserve or increase the amount of carbon sequestered in soil.

Purchase of Carbon Allowances from Emissions Trading Schemes

Voluntary purchasers can offset their carbon emissions by purchasing carbon allowances from legally mandated cap-and-trade programs such as the Regional Greenhouse Gas Initiative or the European Emissions Trading Scheme. By purchasing the allowances that power plants, oil refineries, and industrial facilities need to hold to comply with a cap, voluntary purchases tighten the cap and force additional emissions reductions.

Voluntary purchases can also be made through small-scale and sometimes uncertified schemes such as those offered at South African based Promoting Access to Carbon Equity Centre (PACE), which nevertheless offer clear services such as poverty alleviation in the form of renewable energy development. Also, as "easy carbon credits are coming to an end", these projects have the potential to develop projects that are either too small or too complicated to benefit from legally mandated cap-and-trade programs.

Links with Emission Trading Schemes

Once it has been accredited by the UNFCCC a carbon offset project can be used as carbon credit and linked with official emission trading schemes, such as the European Union Emission Trading Scheme or Kyoto Protocol, as Certified Emission Reductions. European emission allowances for the 2008–2012 second phase were selling for between 21 and 24 Euros per metric ton of CO_2 as of July 2007.

The voluntary Chicago Climate Exchange also includes a carbon offset scheme that allows offset project developers to sell emissions reductions to CCX members who have voluntarily agreed to meet emissions reduction targets.

The Western Climate Initiative, a regional greenhouse gas reduction initiative by states and provinces along the western rim of North America, includes an offset scheme. Likewise, the Regional Greenhouse Gas Initiative, a similar program in the northeastern U.S., includes an offset program. A credit mechanism that uses offsets may be incorporated in proposed schemes such as the Australian Carbon Exchange.

Other

A UK offset provider set up a carbon offsetting scheme that set up a secondary market for treadle pumps in developing countries. These pumps are used by farmers, using human power, in place of diesel pumps. However, given that treadle pumps are best suited to pumping shallow water, while diesel pumps are usually used to pump water from deep boreholes, it is not clear that the treadle pumps are actually achieving real emissions reductions. Other companies have explored and rejected treadle pumps as a viable carbon offsetting approach due to these concerns.

Carbon Retirement

Carbon retirement involves retiring allowances from emission trading schemes as a method for offsetting carbon emissions. Under schemes such as the European Union Emission Trading Scheme, EU Emission Allowances (EUAs), which represent the right to release carbon dioxide into the atmosphere, are issued to all the largest polluters. The theory is that by buying these allowances and permanently removing them, the price of EUAs increases and provides an incentive for industrial companies to reduce their emissions.

Accounting for and Verifying Reductions

Due to their indirect nature, many types of offset are difficult to verify. Some providers obtain independent certification that their offsets are accurately measured, to distance themselves from potentially fraudulent competitors. The credibility of the various certification providers is often questioned. Certified offsets may be purchased from commercial or non-profit organizations for US\$5.50–30 per tonne of CO_2, due to fluctuations of market price. Annual carbon dioxide emissions in developed countries range from 6 to 23 tons per capita.

Accounting systems differ on precisely what constitutes a valid offset for voluntary reduction systems and for mandatory reduction systems. However formal standards for quantification exist based on collaboration between emitters, regulators, environmentalists and project developers. These standards include the Voluntary Carbon Standard, Green-e Climate, Chicago Climate Exchange and the CDM Gold Standard, the latter of which expands upon the requirements for the Clean Development Mechanism of the Kyoto Protocol.

Accounting of offsets may address the following basic areas:

- Baseline and Measurement—What emissions would occur in the absence of a proposed project? And how are the emissions that occur after the project is performed going to be measured?

- Additionality—Would the project occur anyway without the investment raised by selling carbon offset credits? There are two common reasons why a project may lack additionality: (a) if it is intrinsically financially worthwhile due to energy cost savings, and (b) if it had to be performed due to environmental laws or regulations.

- Permanence—Are some benefits of the reductions reversible? (for example, trees may be harvested to burn the wood, and does growing trees for fuel wood decrease the need for fossil fuel?) If woodlands are increasing in area or density, then carbon is being sequestered. After roughly 50 years, newly planted forests will reach maturity and remove carbon dioxide more slowly.

- Leakage—Does implementing the project cause higher emissions outside the project boundary?

Co-benefits

Overall, carbon offsets improve the environment by reducing the amount of greenhouse gases in the earth's atmosphere. Offset projects often also lead to a number of co-benefits such as better air and water quality, and healthier communities.

While the primary goal of carbon offsets is to reduce global carbon emissions, many offset projects also claim to lead to improvements in the quality of life for a local population. These additional improvements are termed *co-benefits*, and may be considered when evaluating and comparing carbon offset projects. Some possible co-benefits from a project that replaces wood-burning stoves with ovens using a less carbon-intensive fuel include:

- Lower non–greenhouse gas pollution (smoke, ash, and chemicals), which improves health in the home.

- Better preservation of forests, which are an important habitat for wildlife.

In a recent survey conducted by EcoSecurities, Conservation International, CCBA and ClimateBiz, of the 120 corporates surveyed more than 77 per cent rated community and environmental benefits as the prime motivator for purchasing carbon offsets.

Carbon offset projects can also negatively affect quality of life. For example, people who earn their livelihoods from collecting firewood and selling it to households could become unemployed if firewood is no longer used. A paper from the Overseas Development Institute offers some indicators to be used in assessing the potential developmental impacts of voluntary carbon offset schemes:

- What potential does the project have for income generation?

- What effects might a project have on future changes in land use and could conflicts arise from this?

- Can small-scale producers engage in the scheme?

- What are the 'add on' benefits to the country—for example, will it assist capacity-building in local institutions?

Putting a price on carbon encourages innovation by providing funding for new ways to reduce greenhouse gases in many sectors. Carbon reduction goals drive the demand for offsets and carbon trading, encouraging the development of this new industry and offering opportunities for different sectors to develop and use innovative new technologies.

Carbon offset projects also provide savings – energy efficiency measures may reduce fuel or electricity consumption, leading to a potential reduction in maintenance and operating costs.

Quality Assurance Schemes

Quality Assurance Standard for Carbon Offsetting (QAS)

In an effort to inform and safeguard business and household consumers purchasing Carbon Offsets, in 2009, the UK Government has launched a scheme for regulating Carbon offset products. DEFRA have created the "Approved Carbon Offsetting" brand to use as an endorsement on offsets approved by the UK government. The Scheme sets standards for best practice in offsetting. Approved offsets have to demonstrate the following criteria:

- Accurate calculation of emissions to be offset
- Use of good quality carbon credits i.e. initially those that are Kyoto compliant
- Cancellation of carbon credits within a year of the consumers purchase of the offset
- Clear and transparent pricing of the offset
- Provision of information about the role of offsetting in tackling climate change and advice on how a consumer can reduce his or her carbon footprint

The first company to qualify for the scheme was Clear, followed by Carbon Footprint, Carbon Passport, Pure, British Airways and Carbon Retirement Ltd.

On 20 May 2011 the Department of Energy and Climate Change announced that the Quality Assurance Scheme would close on 30 June 2011. The stated purpose of the Quality Assurance Scheme was 'to provide a straightforward route for those wishing to offset their emissions to identify quality offsets'. Critics of the closure therefore argued that without the scheme, businesses and individuals would struggle to identify quality carbon offsets.

In 2012 the scheme was relaunched as the Quality Assurance Standard (QAS). The QAS is now run independently by Quality Assurance Standard Ltd which is a company limited by guarantee based in the United Kingdom. The Quality Assurance Standard is a comprehensive independent audit system for carbon offsets. Approved offsets are checked against a 40-point checklist to ensure they meet the very highest standards in the world.

On 17 July 2012, the first organisations were approved as meeting the new QAS.

Australian Government National Carbon Offset Program

The Australian government is currently in a consultation period on the regulation of Carbon Offsets. On 20 December 2013, the Australian Government released the Emissions Reduction Fund Green Paper outlining its preferred design options for the Emissions Reduction Fund: a carbon buy-back model. The Government invites public comment and written submissions on the Green Paper by 5pm on Friday 21 February 2014.

Controversies

Less than 30 pence in every pound spent on some carbon offset schemes goes directly to projects designed to reduce emissions. The figures reported by the BBC and based on UN data reported that typically 28p goes to the set up and maintenance costs of an environmental project. 34p goes to the company that takes on the risk that the project may fail. The project's investors take 19p, with smaller amounts of money being distributed between organisations involved in brokering and auditing the carbon credits. In that respect carbon Offsets are similar to most consumer products, with only a fraction of sale prices going to the off-shore producers, the rest being shared between investors and distributors who bring it to the markets, who themselves need to pay their employees and service providers such as advertising agencies most of the time located in expensive areas.

Indulgence Controversy

Some activists disagree with the principle of carbon offsets, likening them to Roman Catholic indulgences, a way for the guilty to pay for absolution rather than changing their behavior. George Monbiot, an English environmentalist and writer, says that carbon offsets are an excuse for business as usual with regard to pollution. Proponents hold that the indulgence analogy is flawed because they claim carbon offsets actually reduce carbon emissions, changing the business as usual, and therefore address the root cause of climate change. Proponents of offsets claim that third-party certified carbon offsets are leading to increased investment in renewable energy, energy efficiency, methane biodigesters and reforestation and avoided deforestation projects, and claim that these alleged effects are the intended goal of carbon offsets. On October 16, 2009 responsibletravel.com, once a strong voice in favour of carbon offsetting, announced that it would stop offering carbon offsetting to its clients, stating that "too often offsets are being used by the tourism industry in developed countries to justify growth plans on the basis that money will be donated to projects in developing countries. Global reduction targets will not be met this way".

On 4 February 2010, travel networking site Vida Loca Travel announced that they would donate 5 per cent of profits to International Medical Corps, as they feel that international aid can be more effective at cutting global warming in the long term than carbon offsetting, citing the work of economist Jeffrey Sachs.

Effectiveness of Tree-planting Offsets

Some environmentalists have questioned the effectiveness of tree-planting projects for carbon offset purposes. Critics point to the following issues with tree planting projects:

- Timing. Trees reach maturity over a course of many decades. Project developers and offset retailers typically pay for the project and sell the promised reductions up-front, a practice known as "forward selling".

- Permanence. It is difficult to guarantee the permanence of the forests, which may be susceptible to clearing, burning, or mismanagement. The well-publicized instance of the

"Coldplay forest", in which a forestry project supported by the British band Coldplay resulted in a grove of dead mango trees, illustrates the difficulties of guaranteeing the permanence of tree-planting offsets. When discussing "tree offsets, forest campaigner Jutta Kill of European environmental group FERN, clarified the physical reality that "Carbon in trees is temporary: Trees can easily release carbon into the atmosphere through fire, disease, climatic changes, natural decay and timber harvesting."

- Monocultures and invasive species. In an effort to cut costs, some tree-planting projects introduce fast-growing invasive species that end up damaging native forests and reducing biodiversity. For example, in Ecuador, the Dutch FACE Foundation has an offset project in the Andean Páramo involving 220 square kilometres of eucalyptus and pine planted. The NGO Acción Ecológica criticized the project for destroying a valuable Páramo ecosystem by introducing exotic tree species, causing the release of much soil carbon into the atmosphere, and harming local communities who had entered into contracts with the FACE Foundation to plant the trees. However, some certification standards, such as the Climate Community and Biodiversity Standard require multiple species plantings.

- Methane. A recent study has claimed that plants are a significant source of methane, a potent greenhouse gas, raising the possibility that trees and other terrestrial plants may be significant contributors to global methane levels in the atmosphere. However, this claim has been disputed recently by findings in another study.

- The albedo effect. Another study suggested that "high latitude forests probably have a net warming effect on the Earth's climate", because their absorption of sunlight creates a warming effect that balances out their absorption of carbon dioxide.

- Necessity. Corporate tree-planting is not a new idea; farming operations have been used by companies making paper from trees for a long time. If farmed trees are replanted, and the products made from them are placed into landfills rather than recycled, a very safe, efficient, economical and time-proven method of geological sequestration of greenhouse carbon is the result of the paper product use cycle. This only holds if the paper in the land fill is not decomposed. In most landfills, this is the case and leads to the fact that more than half of the greenhouse gas emissions from the life cycle of paper products occur from landfill methane emissions.

Indigenous Land Rights Issues

Tree-planting projects can cause conflicts with indigenous people who are displaced or otherwise find their use of forest resources curtailed. For example, a World Rainforest Movement report documents land disputes and human rights abuses at Mount Elgon. In March 2002, a few days before receiving Forest Stewardship Council certification for a project near Mount Elgon, the Uganda Wildlife Authority evicted more than 300 families from the area and destroyed their homes and crops. That the project was taking place in an area of on-going land conflict and alleged human rights abuses did not make it into project report. A 2011 report by Oxfam International describes a case where over 20,000 farmers in Uganda were displaced for a FSC-certified plantation to offset carbon by London-based New Forests Company.

Additionality and Lack of Regulation in the Voluntary Market

Several certification standards exist, offering variations for measuring emissions baseline, reductions, additionality, and other key criteria. However, no single standard governs the industry, and some offset providers have been criticized on the grounds that carbon reduction claims are exaggerated or misleading. Problems include:

- Widespread instances of people and organizations buying worthless credits that do not yield any reductions in carbon emissions.

- Industrial companies profiting from doing very little – or from gaining carbon credits on the basis of efficiency gains from which they have already benefited substantially.

- Brokers providing services of questionable or no value.

- A shortage of verification, making it difficult for buyers to assess the true value of carbon credits.

Perverse Incentives

Because offsets provide a revenue stream for the reduction of some types of emissions, they can in some cases provide incentives to emit more, so that emitting entities can later get credit for reducing emissions from an artificially high baseline. This is especially the case for offsets with a high profit margin. For example, one Chinese company generated $500 million in carbon offsets by installing a $5 million incinerator to burn the HFCs produced by the manufacture of refrigerants. The huge profits provided incentive to create new factories or expand existing factories solely for the purpose of increasing production of HFCs and then destroying the resultant pollutants to generate offsets. Not only is this outcome environmentally undesirable, it undermines other offset projects by causing offset prices to collapse. The practice had become so common that offset credits are now no longer awarded for new plants to destroy HFC-23.

In Nigeria oil companies *flare off* 40 per cent of the natural gas found. The Agip Oil Company plans to build plants to generate electricity from this gas and thus claim 1.5 million offset credits a year. United States company Pan Ocean Oil Corporation has also applied for credits in exchange for processing its own waste gas in Nigeria. Oilwatch.org's Michael Karikpo calls this "outrageous", as flaring is illegal in Nigeria, adding that *"It's like a criminal demanding money to stop committing crimes"*.

Other Negative Impacts from Offset Projects

Although many carbon offset projects tout their environmental co-benefits, some are accused of having negative secondary effects. Point Carbon has reported on an inconsistent approach with regard to some hydro-electric projects as carbon offsets; some countries in the EU are not allowing large projects into the EU ETS, because of their environmental impacts, even though they have been individually approved by the UNFCCC and World Commission on Dams. It is difficult to assess the exact results of carbon offsets given the fact that they are a relatively new form of carbon reduction, and it is possible that some carbon offset purchases are made in an attempt to increase positive business public relations rather than to help solve the issue of greenhouse gas emissions.

Offset projects may also have negative social impacts, for example when local residents are evicted to enable a National Park to be marketed as a carbon offset.

Emission Trading

Emissions trading or cap and trade is a government-mandated, market-based approach to controlling pollution by providing economic incentives for achieving reductions in the emissions of pollutants. Various countries, states and groups of companies have adopted such trading systems, notably for mitigating climate change.

A central authority (usually a governmental body) allocates or sells a limited number of permits to discharge specific quantities of a specific pollutant per time period. Polluters are required to hold permits in amount equal to their emissions. Polluters that want to increase their emissions must buy permits from others willing to sell them. Financial derivatives of permits can also be traded on second ary markets.

In theory, polluters who can reduce emissions most cheaply will do so, achieving the emission reduction at the lowest cost to society. Cap and trade is meant to provide the private sector with the flexibility required to reduce emissions while stimulating technological innovation and economic growth.

There are active trading programs in several air pollutants. For greenhouse gases, which may cause dangerous climate change, permit units are often called *carbon credits*. The largest greenhouse gases trading program is the European Union Emission Trading Scheme, which trades primarily in *European Union Allowances* (*EUAs*); the Californian scheme trades in California Carbon Allowances, the New Zealand scheme in New Zealand Units and the Australian scheme in Australian Units. The United States has a national market to reduce acid rain and several regional markets in nitrogen oxides.

Overview

A coal power plant in Germany. Due to emissions trading, coal may become a less competitive fuel than other options.

Pollution is the prime example of a market externality. An externality is an effect of some activity on an entity (such as a person) that is not party to a market transaction related to that activity. Emissions trading is a market-based approach, among others, to address pollution. The overall goal of an emissions trading plan is to minimize the cost of meeting a set emissions target.

In an emissions trading system, the government sets an overall limit on emissions, and defines permits (also called allowances), or limited authorizations to emit, up to the level of the overall limit. The government may sell the permits, but in many existing schemes, it gives permits to participants (regulated polluters) equal to each participant's baseline emissions. The baseline is determined by reference to the participant's historical emissions. To demonstrate compliance, a participant must hold permits at least equal to the quantity of pollution it actually emitted during the time period. If every participant complies, the total pollution emitted will be at most equal to the sum of individual limits. Because permits can be bought and sold, a participant can choose either to use its permits exactly (by reducing its own emissions); or to emit less than its permits, and perhaps sell the excess permits; or to emit more than its permits, and buy permits from other participants. In effect, the buyer pays a charge for polluting, while the seller gains a reward for having reduced emissions.

In many schemes, organizations which do not pollute (and therefore have no obligations) may also trade permits and financial derivatives of permits. In some schemes, participants can bank allowances to use in future periods. In some schemes, a proportion of all traded permits must be retired periodically, causing a net reduction in emissions over time. Thus, environmental groups may buy and retire permits, driving up the price of the remaining permits according to the law of demand. In most schemes, permit owners can donate permits to a nonprofit entity and receive a tax deduction. Usually, the government lowers the overall limit over time, with an aim towards a national emissions reduction target.

According to the Environmental Defense Fund, cap-and-trade is the most environmentally and economically sensible approach to controlling greenhouse gas emissions, the primary cause of global warming, because it sets a limit on emissions, and the trading incentivizes companies innovate in order to emit less.

"International trade can offer a range of positive and negative incentives to promote international cooperation on climate change (robust evidence, medium agreement). Three issues are key to developing constructive relationships between international trade and climate agreements: how existing trade policies and rules can be modified to be more climate friendly; whether border adjustment measures (BAMs) or other trade measures can be effective in meeting the goals of international climate agreements; whether the UNFCCC, World Trade Organization (WTO), hybrid of the two, or a new institution is the best forum for a trade-and-climate architecture."

Market and Least-Cost

Economy-wide pricing of carbon is the centre piece of any policy designed to reduce emissions at the lowest possible costs.

Many economists have urged the use of market-based instruments such as emissions trading to address environmental problems instead of prescriptive "command-and-control" regulation. Command-and-control regulation is criticized for being insensitive to geographical and technological differences, and therefore inefficient. After an emissions limit has been set by a government political process, individual companies are free to choose how or whether to reduce their emissions. Failure to report emissions and surrender emission permits is often punishable by a further government regulatory mechanism, such as a fine that increases costs of production. Firms will choose the least-cost way to comply with the pollution regulation, which will lead to reductions where the least expensive solutions exist, while allowing emissions that are more expensive to reduce.

Under an emissions trading system, each regulated polluter has flexibility to use the most cost-effective combination of buying or selling emission permits, reducing its emissions by installing cleaner technology, or reducing its emissions by reducing production. The most cost-effective strategy depends on the polluter's marginal abatement cost and the market price of permits. In theory, a polluter's decisions should lead to an economically efficient allocation of reductions among polluters, and lower compliance costs for individual firms and for the economy overall, compared to command-and-control mechanisms.

Emission Markets

For emissions trading where greenhouse gases are regulated, one emissions permit is considered equivalent to one metric ton of carbon dioxide (CO_2) emissions. Other names for emissions permits are carbon credits, Kyoto units, assigned amount units, and Certified Emission Reduction units (CER). These permits can be sold privately or in the international market at the prevailing market price. These trade and settle internationally, and hence allow permits to be transferred between countries. Each international transfer is validated by the United Nations Framework Convention on Climate Change (UNFCCC). Each transfer of ownership within the European Union is additionally validated by the European Commission.

Emissions trading programmes such as the European Union Emissions Trading System (EU ETS) complement the country-to-country trading stipulated in the Kyoto Protocol by allowing private trading of permits. Under such programmes – which are generally co-ordinated with the national emissions targets provided within the framework of the Kyoto Protocol – a national or international authority allocates permits to individual companies based on established criteria, with a view to meeting national and/or regional Kyoto targets at the lowest overall economic cost.

Trading exchanges have been established to provide a spot market in permits, as well as futures and options market to help discover a market price and maintain liquidity. Carbon prices are normally quoted in euros per tonne of carbon dioxide or its equivalent (CO_2e). Other greenhouse gases can also be traded, but are quoted as standard multiples of carbon dioxide with respect to their global warming potential. These features reduce the quota's financial impact on business, while ensuring that the quotas are met at a national and international level.

Currently, there are six exchanges trading in UNFCCC related carbon credits: the Chicago Climate Exchange (until 2010), European Climate Exchange, NASDAQ OMX Commodities Europe, PowerNext, Commodity Exchange Bratislava and the European Energy Exchange. NASDAQ OMX Commodities Europe listed a contract to trade offsets generated by a CDM carbon project called Certified Emission Reductions. Many companies now engage in emissions abatement, offsetting, and sequestration programs to generate credits that can be sold on one of the exchanges. At least one private electronic market has been established in 2008: CantorCO2e. Carbon credits at Commodity Exchange Bratislava are traded at special platform called Carbon place.

Trading in emission permits is one of the fastest-growing segments in financial services in the City of London with a market estimated to be worth about €30 billion in 2007. Louis Redshaw, head of environmental markets at Barclays Capital, predicts that "carbon will be the world's biggest commodity market, and it could become the world's biggest market overall."

History

The international community began the long process towards building effective international and domestic measures to tackle GHG(Carbon dioxide, methane, nitrous oxide, hydroflurocarbons, perfluorocarbons and sulphur hexafluoride) emissions in response to the increasing assertions that global warming is happening due to man-made emissions and the uncertainty over its likely consequences. That process began in Rio in 1992, when 160 countries agreed the UN Framework Convention on Climate Change (UNFCCC). The UNFCCC is, as its title suggests, simply a frame-work; the necessary detail was left to be settled by the Conference of Parties (CoP) to the UNFCCC.

The efficiency of what later was to be called the "cap-and-trade" approach to air pollution abatement was first demonstrated in a series of micro-economic computer simulation studies between 1967 and 1970 for the National Air Pollution Control Administration (predecessor to the United States Environmental Protection Agency's Office of Air and Radiation) by Ellison Burton and William Sanjour. These studies used mathematical models of several cities and their emission sources in order to compare the cost and effectiveness of various control strategies. Each abatement strategy was compared with the "least cost solution" produced by a computer optimization program to identify the least costly combination of source reductions in order to achieve a given abatement goal. In each case it was found that the least cost solution was dramatically less costly than the same amount of pollution reduction produced by any conventional abatement strategy. Burton and later Sanjour along with Edward H. Pechan continued improving and advancing these computer models at the newly created U.S. Environmental Protection Agency. The agency introduced the concept of computer modeling with least cost abatement strategies (i.e. emissions trading) in its 1972 annual report to Congress on the cost of clean air. This led to the concept of "cap and trade" as a means of achieving the "least cost solution" for a given level of abatement.

The development of emissions trading over the course of its history can be divided into four phases:

1. Gestation: Theoretical articulation of the instrument (by Coase, Crocker, Dales, Montgomery etc.) and, independent of the former, tinkering with "flexible regulation" at the US Environmental Protection Agency.

2. Proof of Principle: First developments towards trading of emission certificates based on the "offset-mechanism" taken up in Clean Air Act in 1977. A company could get allowance from the Act on greater amount of emission when it paid another company to reduce the same pollutant.

3. Prototype: Launching of a first "cap-and-trade" system as part of the US Acid Rain Program in Title IV of the 1990 Clean Air Act, officially announced as a paradigm shift in environmental policy, as prepared by "Project 88", a network-building effort to bring together environmental and industrial interests in the US.

4. Regime formation: branching out from the US clean air policy to global climate policy, and from there to the European Union, along with the expectation of an emerging global carbon market and the formation of the "carbon industry".

In the United States, the "acid rain"-related emission trading system was principally conceived by C. Boyden Gray, a G.H.W. Bush administration attorney. Gray worked with the Environmental Defense Fund (EDF), who worked with the EPA to write the bill that became law as part of the Clean Air Act of 1990. The new emissions cap on NO_x and SO_2 gases took effect in 1995, and according to *Smithsonian* magazine, those acid rain emissions dropped 3 million tons that year. In 1997, the CoP agreed, in what has been described as a watershed in international environmental treaty making, the Kyoto Protocol where 38 developed countries(Annex 1 countries.) committed themselves to targets and timetables for the reduction of GHGs. These targets for developed countries are often referred to as Assigned Amounts.

One important economic reality recognised by many of the countries that signed the Kyoto Protocol is that, if countries have to solely rely on their own domestic measures, the resulting inflexible limitations on GHG growth could entail very large costs, perhaps running into many trillions of dollars globally. As a result, international mechanisms which would allow developed countries flexibility to meet their targets were included in the Kyoto Protocol. The purpose of these mechanisms is to allow the parties to find the most economic ways to achieve their targets. These international mechanisms are outlined under Kyoto Protocol.

On April 17, 2009, the Environmental Protection Agency (EPA) formally announced that it had found that greenhouse gas (GHG) poses a threat to public health and the environment (EPA 2009a). This announcement was significant because it gives the executive branch the authority to impose carbon regulations on carbon-emitting entities.

A carbon cap-and-trade system is to be introduced nationwide in China in 2016 (China's National Development and Reform Commission proposed that an absolute cap be placed on emission by 2016.)

Public Opinion

In the United States, most polling shows large support for emissions trading (often referred to as cap-and-trade). This majority support can be seen in polls conducted by Washington Post/ABC News, Zogby International and Yale University. A new Washington Post-ABC poll reveals that majorities of the American people believe in climate change, are concerned about it, are willing to change their lifestyles and pay more to address it, and want the federal government to regulate greenhouse gases. They are, however, ambivalent on cap-and-trade.

More than three-quarters of respondents, 77.0%, reported they "strongly support" (51.0%) or "somewhat support" (26.0%) the EPA's decision to regulate carbon emissions. While 68.6% of respondents reported being "very willing" (23.0%) or "somewhat willing" (45.6%), another 26.8% reported being "somewhat unwilling" (8.8%) or "not at all willing" (18.0%) to pay higher prices for "Green" energy sources to support funding for programs that reduce the effect of global warming.

According to PolitiFact, it is a misconception that emissions trading is unpopular in the United

States because of earlier polls from Zogby International and Rasmussen which misleadingly include "new taxes" in the questions (taxes aren't part of emissions trading) or high energy cost estimates.

Comparison with other Methods of Emission Reduction

Cap and trade is the textbook emissions trading program. Other market-based approaches include baseline-and-credit, and pollution tax. They all put a price on pollution, and so provide an economic incentive to reduce pollution beginning with the lowest-cost opportunities. By contrast, in a command-and-control approach, a central authority designates pollution levels each facility is allowed to emit.

Baseline and Credit

In a baseline and credit program, polluters can create permits, called credits or offsets, by reducing their emissions below a baseline level, which is often the historical emissions level from a designated past year. Such credits can be bought by polluters that have a regulatory limit.

Pollution Tax

Emissions fees or environmental tax is a surcharge on the pollution created while producing goods and services. For example, a carbon tax is a tax on the carbon content of fossil fuels that aims to discourage their use and thereby reduce carbon dioxide emissions. The two approaches are overlapping sets of policy designs. Both can have a range of scopes, points of regulation, and price schedules. They can be fair or unfair, depending on how the revenue is used. Both have the effect of increasing the price of goods (such as fossil fuels) to consumers. A comprehensive, upstream, auctioned cap-and-trade system is very similar to a comprehensive, upstream carbon tax. Yet, many commentators sharply contrast the two approaches.

The main difference is what is defined and what derived. A tax is a price control, while cap-and-trade method acts is a quantity control instrument. That is, a tax is a unit price for pollution that is set by authorities, and the market determines the quantity emitted; in cap and trade, authorities determine the amount of pollution, and the market determines the price. This difference affects a number of criteria.

Responsiveness to inflation: Cap-and-trade has the advantage that it adjusts to inflation (changes to overall prices) automatically, while emissions fees must be changed by regulators.

Responsiveness to cost changes: It is not clear which approach is better. It is possible to combine the two into a safety valve price: a price set by regulators, at which polluters can buy additional permits beyond the cap.

Responsiveness to recessions: This point is closely related to responsiveness to cost changes, because recessions cause a drop in demand. Under cap and trade, the emissions cost automatically decreases, so a cap-and-trade scheme adds another automatic stabilizer to the economy - in effect, an automatic fiscal stimulus. However, a lower pollution price also results in reduced efforts to reduce pollution. If the government is able to stimulate the economy regardless of the cap-and-trade scheme, an excessively low price causes a missed opportunity to cut emissions faster than planned. Instead, it might be better to have a price floor (a tax). This is especially true when cutting pollution is urgent, as with greenhouse gas emissions. A price floor also provides certainty and stability for investment in emissions reductions: recent experience from the UK shows that nuclear power operators are reluctant to invest on "un-subsidised" terms unless there is a guaranteed price floor for carbon (which the EU emissions trading scheme does not presently provide).

Responsiveness to uncertainty: As with cost changes, in a world of uncertainty, it is not clear whether emissions fees or cap-and-trade systems are more efficient—it depends on how fast the marginal social benefits of reducing pollution fall with the amount of cleanup (e.g., whether inelastic or elastic marginal social benefit schedule).

Other: The magnitude of the tax will depend on how sensitive the supply of emissions is to the price. The permit price of cap-and-trade will depend on the pollutant market. A tax generates government revenue, but full-auctioned emissions permits can do the same. A similar upstream cap-and-trade system could be implemented. An upstream carbon tax might be the simplest to administer. Setting up a complex cap-and-trade arrangement that is comprehensive has high institutional needs.

Command-and-Control Regulation

Command and control is a system of regulation that prescribes emission limits and compliance methods for each facility or source. It is the traditional approach to reducing air pollution.

Command-and-control regulations are more rigid than incentive-based approaches such as pollution fees and cap and trade. An example of this is a performance standard which sets an emissions goal for each polluter that is fixed and, therefore, the burden of reducing pollution cannot be shifted to the firms that can achieve it more cheaply. As a result, performance standards are likely to be more costly overall. The additional costs would be passed to end consumers.

Economics of International Emissions Trading

It is possible for a country to reduce emissions using a Command-Control approach, such as regulation, direct and indirect taxes. The cost of that approach differs between countries because the Marginal Abatement Cost Curve (MAC) — the cost of eliminating an additional unit of pollution — differs by country. It might cost China $2 to eliminate a ton of CO_2, but it would probably cost Norway or the U.S. much more. International emissions-trading markets were created precisely to exploit differing MACs.

Example

Emissions trading through *Gains from Trade* can be more beneficial for both the buyer and the seller than a simple emissions capping scheme.

Consider two European countries, such as Germany and Sweden. Each can either reduce all the required amount of emissions by itself or it can choose to buy or sell in the market.

Emissions Trading

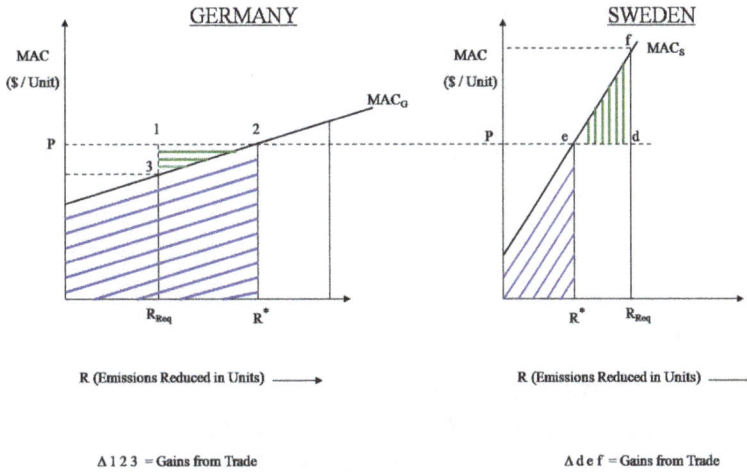

Example MACs for two different countries

Suppose Germany can abate its CO_2 at a much cheaper cost than Sweden, i.e. $MAC_S > MAC_G$ where the MAC curve of Sweden is steeper (higher slope) than that of Germany, and R_{Req} is the total amount of emissions that need to be reduced by a country.

On the left side of the graph is the MAC curve for Germany. R_{Req} is the amount of required reductions for Germany, but at R_{Req} the MAC_G curve has not intersected the market emissions permit price of CO_2 (market permit price = P = λ). Thus, given the market price of CO_2 allowances, Germany has potential to profit if it abates more emissions than required.

On the right side is the MAC curve for Sweden. R_{Req} is the amount of required reductions for Sweden, but the MAC_S curve already intersects the market price of CO_2 permits before R_{Req} has been reached. Thus, given the market price of CO_2 permits, Sweden has potential to make a cost saving if it abates fewer emissions than required internally, and instead abates them elsewhere.

In this example, Sweden would abate emissions until its MAC_S intersects with P (at R*), but this would only reduce a fraction of Sweden's total required abatement.

After that it could buy emissions credits from Germany for the price P (per unit). The internal cost of Sweden's own abatement, combined with the permits it buys in the market from Germany, adds up to the total required reductions (R_{Req}) for Sweden. Thus Sweden can make a saving from buying permits in the market (Δ d-e-f). This represents the "Gains from Trade", the amount of additional expense that Sweden would otherwise have to spend if it abated all of its required emissions by itself without trading.

Germany made a profit on its additional emissions abatement, above what was required: it met the regulations by abating all of the emissions that was required of it (R_{Req}). Additionally, Germany sold its surplus permits to Sweden, and was paid P for every unit it abated, while spending less than P. Its total revenue is the area of the graph (R_{Req} 1 2 R*), its total abatement cost is area (R_{Req} 3 2 R*), and so its net benefit from selling emission permits is the area (Δ 1-2-3) i.e. Gains from Trade

The two R* (on both graphs) represent the efficient allocations that arise from trading.

- Germany: sold (R* - R$_{Req}$) emission permits to Sweden at a unit price P.

- Sweden bought emission permits from Germany at a unit price P.

If the total cost for reducing a particular amount of emissions in the *Command Control* scenario is called X, then to reduce the same amount of combined pollution in Sweden and Germany, the total abatement cost would be less in the *Emissions Trading* scenario i.e. (X − Δ 123 - Δ def).

The example above applies not just at the national level, but also between two companies in different countries, or between two subsidiaries within the same company.

Applying the Economic Theory

The nature of the pollutant plays a very important role when policy-makers decide which framework should be used to control pollution. CO_2 acts globally, thus its impact on the environment is generally similar wherever in the globe it is released. So the location of the originator of the emissions does not matter from an environmental standpoint.

The policy framework should be different for regional pollutants (e.g. SO_2 and NO_x, and also mercury) because the impact of these pollutants may differ by location. The same amount of a regional pollutant can exert a very high impact in some locations and a low impact in other locations, so it matters where the pollutant is released. This is known as the *Hot Spot* problem.

A Lagrange framework is commonly used to determine the least cost of achieving an objective, in this case the total reduction in emissions required in a year. In some cases, it is possible to use the Lagrange optimization framework to determine the required reductions for each country (based on their MAC) so that the total cost of reduction is minimized. In such a scenario, the Lagrange multiplier represents the market allowance price (P) of a pollutant, such as the current market price of emission permits in Europe and the USA.

Countries face the permit market price that exists in the market that day, so they are able to make individual decisions that would minimize their costs while at the same time achieving regulatory compliance. This is also another version of the Equi-Marginal Principle, commonly used in economics to choose the most economically efficient decision.

Prices Versus Quantities, and the Safety Valve

There has been longstanding debate on the relative merits of *price* versus *quantity* instruments to achieve emission reductions.

An emission cap and permit trading system is a *quantity* instrument because it fixes the overall emission level (quantity) and allows the price to vary. Uncertainty in future supply and demand conditions (market volatility) coupled with a fixed number of pollution permits creates an uncertainty in the future price of pollution permits, and the industry must accordingly bear the cost of adapting to these volatile market conditions. The burden of a volatile market thus lies with the industry rather than the controlling agency, which is generally more efficient. However, under volatile market conditions, the ability of the controlling agency to alter the caps will translate into an ability to pick "winners and losers" and thus presents an opportunity for corruption.

In contrast, an emission tax is a *price* instrument because it fixes the price while the emission level is allowed to vary according to economic activity. A major drawback of an emission tax is that the environmental outcome (e.g. a limit on the amount of emissions) is not guaranteed. On one hand, a tax will remove capital from the industry, suppressing possibly useful economic activity, but conversely, the polluter will not need to hedge as much against future uncertainty since the amount of tax will track with profits. The burden of a volatile market will be borne by the controlling (taxing) agency rather than the industry itself, which is generally less efficient. An advantage is that, given a uniform tax rate and a volatile market, the taxing entity will not be in a position to pick "winners and losers" and the opportunity for corruption will be less.

Assuming no corruption and assuming that the controlling agency and the industry are equally efficient at adapting to volatile market conditions, the best choice depends on the sensitivity of the costs of emission reduction, compared to the sensitivity of the benefits (i.e., climate damage avoided by a reduction) when the level of emission control is varied.

Because there is high uncertainty in the compliance costs of firms, some argue that the optimum choice is the price mechanism. However, the burden of uncertainty cannot be eliminated, and in this case it is shifted to the taxing agency itself.

Some scientists have warned of a threshold in atmospheric concentrations of carbon dioxide beyond which a run-away warming effect could take place, with a large possibility of causing irreversible damage. With such a risk, a quantity instrument may be a better choice because the quantity of emissions may be capped with more certainty. However, this may not be true if this risk exists but cannot be attached to a known level of greenhouse gas (GHG) concentration or a known emission pathway.

A third option, known as a *safety valve*, is a hybrid of the price and quantity instruments. The system is essentially an emission cap and permit trading system but the maximum (or minimum) permit price is capped. Emitters have the choice of either obtaining permits in the marketplace or buying them from the government at a specified trigger price (which could be adjusted over time). The system is sometimes recommended as a way of overcoming the fundamental disadvantages of both systems by giving governments the flexibility to adjust the system as new information comes to light. It can be shown that by setting the trigger price high enough, or the number of permits low enough, the safety valve can be used to mimic either a pure quantity or pure price mechanism.

All three methods are being used as policy instruments to control greenhouse gas emissions: the EU-ETS is a *quantity* system using the cap and trading system to meet targets set by National Allocation Plans; Denmark has a price system using a carbon tax, while China uses the CO_2 market price for funding of its Clean Development Mechanism projects, but imposes a *safety valve* of a minimum price per tonne of CO_2.

Carbon Leakage

Carbon leakage is the effect that regulation of emissions in one country/sector has on the emissions in other countries/sectors that are not subject to the same regulation. There is no consensus over the magnitude of long-term carbon leakage.

In the Kyoto Protocol, Annex I countries are subject to caps on emissions, but non-Annex I countries are not. Barker *et al.* (2007) assessed the literature on leakage. The leakage rate is defined as the increase in CO_2 emissions outside the countries taking domestic mitigation action, divided by the reduction in emissions of countries taking domestic mitigation action. Accordingly, a leakage rate greater than 100% means that actions to reduce emissions within countries had the effect of increasing emissions in other countries to a greater extent, i.e., domestic mitigation action had actually led to an increase in global emissions.

Estimates of leakage rates for action under the Kyoto Protocol ranged from 5% to 20% as a result of a loss in price competitiveness, but these leakage rates were considered very uncertain. For energy-intensive industries, the beneficial effects of Annex I actions through technological development were considered possibly substantial. However, this beneficial effect had not been reliably quantified. On the empirical evidence they assessed, Barker *et al.* (2007) concluded that the competitive losses of then-current mitigation actions, e.g., the EU ETS, were not significant.

Under the EU ETS rules Carbon Leakage Exposure Factor is used to determine the volumes of free allocation of emission permits to industrial installations.

Trade

To understand carbon trading, it is important to understand the products that are being traded. The primary product in carbon markets is the trading of GHG emission permits. Under a cap-and-trade system, permits are issued to various entities for the right to emit GHG emissions that meet emission reduction requirement caps.

One of the controversies about carbon mitigation policy is how to "level the playing field" with border adjustments. For example, one component of the American Clean Energy and Security Act, along with several other energy bills put before US Congress, calls for carbon surcharges on goods imported from countries without cap-and-trade programs. Besides issues of compliance with the General Agreement on Tariffs and Trade, such border adjustments presume that the producing countries bear responsibility for the carbon emissions.

A general perception among developing countries is that discussion of climate change in trade negotiations could lead to "green protectionism" by high-income countries (World Bank, 2010, p. 251). Tariffs on imports ("virtual carbon") consistent with a carbon price of $50 per ton of CO_2 could be significant for developing countries. World Bank (2010) commented that introducing border tariffs could lead to a proliferation of trade measures where the competitive playing field is viewed as being uneven. Tariffs could also be a burden on low-income countries that have contributed very little to the problem of climate change.

Trading Systems

Kyoto Protocol

As the Intergovernmental Panel on Climate Change (IPCC) reports came in over the years, they shed abundant light on the true state of global warming and they gave support to the environmental effort to address this unprecedented problem. However, the same discussions that started decades back had never ceased and the crusade for a tangible solution to global climate change had gone on all the while. In 1997 the Kyoto Protocol was adopted. The Kyoto Protocol is a 1997 international treaty that came into force in 2005. In the treaty, most developed nations agreed to legally binding targets for their emissions of the six major greenhouse gases. Emission quotas (known as "Assigned amounts") were agreed by each participating 'Annex I' country, with the intention of reducing the overall emissions by 5.2% from their 1990 levels by the end of 2012. The United States is the only industrialized nation under Annex I that has not ratified the treaty, and is therefore not bound by it. The IPCC has projected that the financial effect of compliance through trading within the Kyoto commitment period will be limited at between 0.1-1.1% of GDP among trading countries. The agreement was intended to result in industrialized countries' emissions declining in aggregate by 5.2 percent below 1990 levels by the year of 2012. Despite the failure of the United States and Australia to ratify the protocol, the agreement became effective in 2005, once the requirement that 55 Annex I (predominantly industrialized) countries, jointly accounting for 55 percent of 1990 Annex I emissions, ratify the agreement was met.

The Protocol defines several mechanisms ("flexible mechanisms") that are designed to allow Annex I countries to meet their emission reduction commitments (caps) with reduced economic impact (IPCC, 2007).

Under Article 3.3 of the Kyoto Protocol, Annex I Parties may use GHG removals, from afforestation and reforestation (forest sinks) and deforestation (sources) since 1990, to meet their emission reduction commitments.

Annex I Parties may also use International Emissions Trading (IET). Under the treaty, for the 5-year compliance period from 2008 until 2012, nations that emit less than their quota will be able to sell assigned amount units (each AAU representing an allowance to emit one metric tonne of CO_2) to nations that exceed their quotas. It is also possible for Annex I countries to sponsor carbon projects that reduce greenhouse gas emissions in other countries. These projects generate tradable carbon credits that can be used by Annex I countries in meeting their caps. The project-based Kyoto Mechanisms are the Clean Development Mechanism (CDM) and Joint Implementation (JI). There are four such international flexible mechanisms, or Kyoto Mechanism written in the Kyoto Protocol.

Article 17 if the Protocol authorizes Annex 1 countries that have agreed to the emissions limitations to take part in emissions trading with other Annex 1 Countries.

Article 4 authorizes such parties to implement their limitations jointly, as the member states of the EU have chosen to do.

Article 6 provides that such Annex 1 countries may take part in joint initiatives (JIs) in return for emissions reduction units (ERUs) to be used against their Assigned Amounts.

Art 12 provides for a mechanism known as the clean development mechanism (CDM), under which Annex 1 countries may invest in emissions limitation projects in developing countries and use certified emissions reductions (CERs) generated against their own Assigned Amounts.

The CDM covers projects taking place in non-Annex I countries, while JI covers projects taking place in Annex I countries. CDM projects are supposed to contribute to sustainable development in developing countries, and also generate "real" and "additional" emission savings, i.e., savings that only occur thanks to the CDM project in question (Carbon Trust, 2009, p. 14). Whether or not these emission savings are genuine is, however, difficult to prove.

Australia

In 2003 the New South Wales (NSW) state government unilaterally established the NSW Greenhouse Gas Abatement Scheme to reduce emissions by requiring electricity generators and large consumers to purchase NSW Greenhouse Abatement Certificates (NGACs). This has prompted the rollout of free energy-efficient compact fluorescent lightbulbs and other energy-efficiency measures, funded by the credits. This scheme has been criticised by the Centre for Energy and Environmental Markets (CEEM) of the UNSW because of its lack of effectiveness in reducing emissions, its lack of transparency and its lack of verification of the additionality of emission reductions.

Both the incumbent Howard Coalition government and the Rudd Labor opposition promised to implement an emissions trading scheme (ETS) before the 2007 federal election. Labor won the election, with the new government proceeding to implement an ETS. The government introduced the Carbon Pollution Reduction Scheme, which the Liberals supported with Malcolm Turnbull as leader. Tony Abbott questioned an ETS, saying the best way to reduce emissions is with a "simple tax". Shortly before the carbon vote, Abbott defeated Turnbull in a leadership challenge, and from there on the Liberals opposed the ETS. This left the government unable to secure passage of the bill and it was subsequently withdrawn.

Julia Gillard defeated Rudd in a leadership challenge and promised not to introduce a carbon tax, but would look to legislate a price on carbon when taking the government to the 2010 election. In the first hung parliament result in 70 years, the government required the support of cross-benchers including the Greens. One requirement for Greens support was a carbon price, which Gillard proceeded with in forming a minority government. A fixed carbon price would proceed to a floating-price ETS within a few years under the plan. The fixed price leant itself to characterisation as a carbon tax and when the government proposed the Clean Energy Bill in February 2011, the opposition claimed it to be a broken election promise.

The bill was passed by the Lower House in October 2011 and the Upper House in November 2011. The Liberal Party vowed to overturn the bill if elected.

The Liberal/National coalition government elected in September 2013 has promised to reverse the climate legislation of the previous government. In July 2014, the carbon tax was repealed as well as the Emissions Trading Scheme (ETS) that was to start in 2015.

New Zealand

New Zealand Unit Prices 2010 to 2015

The New Zealand Emissions Trading Scheme (NZ ETS) is a partial-coverage all-free allocation uncapped highly internationally linked emissions trading scheme. The NZ ETS was first legislated in the Climate Change Response (Emissions Trading) Amendment Act 2008 in September 2008 under the Fifth Labour Government of New Zealand and then amended in November 2009 and in November 2012 by the Fifth National Government of New Zealand.

The NZ ETS covers forestry (a net sink), energy (43.4% of total 2010 emissions), industry (6.7% of total 2010 emissions) and waste (2.8% of total 2010 emissions) but not pastoral agriculture (47% of 2010 total emissions). Participants in the NZ ETS must surrender one emission unit (either an international 'Kyoto' unit or a New Zealand-issued unit) for every two tonnes of carbon dioxide equivalent emissions reported or they may choose to buy NZ units from the government at a fixed price of NZ$25.

Individual sectors of the economy have different entry dates when their obligations to report emissions and surrender emission units take effect. Forestry, which contributed net removals of 17.5 Mts of CO_2e in 2010 (19% of NZ's 2008 emissions,) entered the NZ ETS on 1 January 2008. The stationary energy, industrial processes and liquid fossil fuel sectors entered the NZ ETS on 1 July 2010. The waste sector (landfill operators) entered on 1 January 2013. Methane and nitrous oxide emissions from pastoral agriculture are not included in the NZ ETS. (From November 2009, agriculture was to enter the NZ ETS on 1 January 2015)

The NZ ETS is highly linked to international carbon markets as it allows the importing of most of the Kyoto Protocol emission units. However, as of June 2015, the scheme will effectively transition into a domestic scheme, with restricted access to international Kyoto units (CERs, ERUs and RMUs). The NZ ETS has a domestic unit; the 'New Zealand Unit' (NZU), which is issued by free allocation to emitters, with no auctions intended in the short term. Free allocation of NZUs varies between sectors. The commercial fishery sector (who are not participants) have a free allocation of units on a historic basis. Owners of pre-1990 forests have received a fixed free allocation of units. Free allocation to emissions-intensive industry, is provided on an output-intensity basis. For this sector, there is no set limit on the number of units that may be allocated. The number of units allocated to eligible emitters is based on the average emissions per unit of output within a defined 'activity'. Bertram and Terry (2010, p 16) state that as the NZ ETS does not 'cap' emissions, the NZ ETS is not a cap and trade scheme as understood in the economics literature.

Some stakeholders have criticized the New Zealand Emissions Trading Scheme for its generous free allocations of emission units and the lack of a carbon price signal (the Parliamentary Commissioner for the Environment), and for being ineffective in reducing emissions (Greenpeace Aotearoa New Zealand).

The NZ ETS was reviewed in late 2011 by an independent panel, which reported to the Government and public in September 2011.

European Union

The European Union Emission Trading Scheme (or EU ETS) is the largest multi-national, greenhouse gas emissions trading scheme in the world. It is one of the EU's central policy instruments to meet their cap set in the Kyoto Protocol.

After voluntary trials in the UK and Denmark, Phase I began operation in January 2005 with all 15 member states of the European Union participating. The program caps the amount of carbon dioxide that can be emitted from large installations with a net heat supply in excess of 20 MW, such as power plants and carbon intensive factories and covers almost half (46%) of the EU's Carbon Dioxide emissions. Phase I permits participants to trade among themselves and in validated credits from the developing world through Kyoto's Clean Development Mechanism. Credits are gained by investing in clean technologies and low-carbon solutions, and by certain types of emission-saving projects around the world to cover a proportion of their emissions.

During Phases I and II, allowances for emissions have typically been given free to firms, which has resulted in them getting windfall profits (CCC, 2008, p. 149). Ellerman and Buchner (2008) (referenced by Grubb et al.., 2009, p. 11) suggested that during its first two years in operation, the EU ETS turned an expected increase in emissions of 1%-2% per year into a small absolute decline. Grubb et al.. (2009, p. 11) suggested that a reasonable estimate for the emissions cut achieved during its first two years of operation was 50-100 MtCO$_2$ per year, or 2.5%-5%.

A number of design flaws have limited the effectiveness of scheme (Jones et al.., 2007, p. 64). In the initial 2005-07 period, emission caps were not tight enough to drive a significant reduction in emissions (CCC, 2008, p. 149). The total allocation of allowances turned out to exceed actual emissions. This drove the carbon price down to zero in 2007. This oversupply was caused because the allocation of allowances by the EU was based on emissions data from the European Environmental Agency in Copenhagen, which uses a horizontal activity-based emissions definition similar to the United Nations, the EU ETS Transaction log in Brussels, but a vertical installation-based emissions measurement system. This caused an oversupply of 200 million tonnes (10% of market) in the EU ETS in the first phase and collapsing prices.

Phase II saw some tightening, but the use of JI and CDM offsets was allowed, with the result that no reductions in the EU will be required to meet the Phase II cap (CCC, 2008, pp. 145, 149). For Phase II, the cap is expected to result in an emissions reduction in 2010 of about 2.4% compared to expected emissions without the cap (business-as-usual emissions) (Jones et al.., 2007, p. 64). For Phase III (2013–20), the European Commission proposed a number of changes, including:

- Setting an overall EU cap, with allowances then allocated to EU members;

- Tighter limits on the use of offsets;

- Unlimited banking of allowances between Phases II and III;

- A move from allowances to auctioning.

In January 2008, Norway, Iceland, and Liechtenstein joined the European Union Emissions Trading System (EU ETS), according to a publication from the European Commission. The Norwegian Ministry of the Environment has also released its draft National Allocation Plan which provides a carbon cap-and-trade of 15 million metric tonnes of CO$_2$, 8 million of which are set to be auctioned. According to the OECD Economic Survey of Norway 2010, the nation "has announced a target for 2008-12 10% below its commitment under the Kyoto Protocol and a 30% cut compared with 1990 by 2020." In 2012, EU-15 emissions was 15.1% below their base year level. Based on figures for 2012 by the European Environment Agency, EU-15 emissions averaged 11.8% below base-year levels during the 2008-2012 period. This means the EU-15 over-achieved its first Kyoto target by a wide margin.

Tokyo, Japan

The Japanese city of Tokyo is like a country in its own right in terms of its energy consumption and GDP. Tokyo consumes as much energy as "entire countries in Northern Europe, and its production matches the GNP of the world's 16th largest country". A scheme to limit carbon emissions launched in April 2010 covers the top 1,400 emitters in Tokyo, and is enforced and overseen by the Tokyo Metropolitan Government. Phase 1, which is similar to Japan's scheme, ran until 2015. (Japan had an ineffective voluntary emissions reductions system for years, but no nationwide cap-and-trade program.) Emitters must cut their emissions by 6% or 8% depending on the type of organization; from 2011, those who exceed their limits must buy matching allowances or invest in renewable-energy certificates or offset credits issued by smaller businesses or branch offices. Polluters that fail to comply will be fined up to 500,000 yen plus credits for 1.3 times excess emissions. In its fourth year, emissions were reduced by 23% compared to base-year emissions. In phase 2, (FY2015-FY2019), the target is expected to increase to 15%-17%. The aim is to cut Tokyo's carbon emissions by 25% from 2000 levels by 2020. These emission limits can be met by using technologies such as solar panels and advanced fuel-saving devices.

United States

An early example of an emission trading system has been the SO_2 trading system under the framework of the Acid Rain Program of the 1990 Clean Air Act in the U.S. Under the program, which is essentially a cap-and-trade emissions trading system, SO_2 emissions were reduced by 50% from 1980 levels by 2007. Some experts argue that the cap-and-trade system of SO_2 emissions reduction has reduced the cost of controlling acid rain by as much as 80% versus source-by-source reduction. The SO_2 program was challenged in 2004, which set in motion a series of events that led to the 2011 Cross-State Air Pollution Rule (CSAPR). Under the CSAPR, the national SO_2 trading program was replaced by four separate trading groups for SO_2 and NO_x.

In 1997, the State of Illinois adopted a trading program for volatile organic compounds in most of the Chicago area, called the Emissions Reduction Market System. Beginning in 2000, over 100 major sources of pollution in eight Illinois counties began trading pollution credits.

In 2003, New York State proposed and attained commitments from nine Northeast states to form a cap-and-trade carbon dioxide emissions program for power generators, called the Regional Greenhouse Gas Initiative (RGGI). This program launched on January 1, 2009 with the aim to reduce the carbon "budget" of each state's electricity generation sector to 10% below their 2009 allowances by 2018.

Also in 2003, U.S. corporations were able to trade CO_2 emission allowances on the Chicago Climate Exchange under a voluntary scheme. In August 2007, the Exchange announced a mechanism to create emission offsets for projects within the United States that cleanly destroy ozone-depleting substances.

Also in 2003, the Environmental Protection Agency (EPA) began to administer the NOx Budget Trading Program (NBP)under the NOx State Implementation Plan (also known as the "NOx SIP Call") The NOx Budget Trading Program was a market-based cap and trade program created to reduce emissions of nitrogen oxides (NO_x) from power plants and other large combustion sources in the eastern United States. NO_x is a prime ingredient in the formation of ground-level ozone (smog), a pervasive air pollution problem in many areas of the eastern United States. The NBP was designed to reduce NO_x emissions during the warm summer months, referred to as the ozone season, when ground-level ozone concentrations are highest. In March 2008, EPA again strengthened the 8-hour ozone standard to 0.075 parts per million (ppm) from its previous 0.008 ppm.

In 2006, the California Legislature passed the California Global Warming Solutions Act, AB-32, which was signed into law by Governor Arnold Schwarzenegger. Thus far, flexible mechanisms in the form of project based offsets have been suggested for three main project types. The project types include: manure management, forestry, and destruction of ozone-depleted substances. However, a recent ruling from Judge Ernest H. Goldsmith of San Francisco's Superior Court states that the rules governing California's cap-and-trade system were adopted without a proper analysis of alternative methods to reduce greenhouse gas emissions. The tentative ruling, issued on January 24, 2011, argues that the California Air Resources Board violated state environmental law by failing to consider such alternatives. If the decision is made final, the state would not be allowed to implement its proposed cap-and-trade system until the California Air Resources Board fully complies with the California Environmental Quality Act.

In February 2007, five U.S. states and four Canadian provinces joined together to create the Western Climate Initiative (WCI), a regional greenhouse gas emissions trading system. In July 2010, a meeting took place to further outline the cap-and-trade system. In November 2011, Arizona, Montana, New Mexico, Oregon, Utah and Washington withdrew from the WCI.

On November 17, 2008 President-elect Barack Obama clarified, in a talk recorded for YouTube, his intentions for the US to enter a cap-and-trade system to limit global warming.

SO_2 emissions from Acid Rain Program sources have fallen from 17.3 million tons in 1980 to about 7.6 million tons in 2008, a decrease in emissions of 56 percent. Ozone season NOx emissions decreased by 43 percent between 2003 and 2008, even while energy demand remained essentially flat during the same period. CAIR will result in $85 billion to $100 billion in health benefits and nearly $2 billion in visibility benefits per year by 2015 and will substantially reduce premature mortality in the eastern United States.A recent EPA analysis shows that implementation of the Acid Rain Program is expected to reduce between 20,000 and 50,000 incidences of premature mortality annually due to reductions of ambient PM2.5 concentrations, and between 430 and 2,000 incidences annually due to reductions of ground-level ozone. NOx reductions due to the NOx Budget Trading Program have led to improvements in ozone and PM2.5, saving an estimated 580 to 1,800 lives in 2008.

The 2010 United States federal budget proposes to support clean energy development with a 10-year investment of US $15 billion per year, generated from the sale of greenhouse gas (GHG) emissions credits. Under the proposed cap-and-trade program, all GHG emissions credits would be auctioned off, generating an estimated $78.7 billion in additional revenue in FY 2012, steadily increasing to $83 billion by FY 2019.

The American Clean Energy and Security Act (H.R. 2454), a greenhouse gas cap-and-trade bill, was passed on June 26, 2009, in the House of Representatives by a vote of 219-212. The bill originated in the House Energy and Commerce Committee and was introduced by Representatives Henry A. Waxman and Edward J. Markey. The political advocacy organizations FreedomWorks and Americans for Prosperity, funded by brothers David and Charles Koch of Koch Industries, encouraged the Tea Party movement to focus on defeating the legislation. Although cap and trade also gained a significant foothold in the Senate via the efforts of Republican Lindsey Graham, Independent and former Democrat Joe Lieberman, and Democrat John Kerry, the legislation died in the Senate.

South Korea

South Korea's national emissions trading scheme officially launched on 1 January 2015, covering 525 entities from 23 sectors. With a three-year cap of 1.8687 billion tCO2e, it now forms the second largest carbon market in the world following the EU ETS.This amounts to roughly two thirds of the country's emissions. The Korean emissions trading scheme is part of the Republic of Korea's efforts to reduce greenhouse gas emissions by 30% compared to the business-as-usual scenario by 2020.

China

In November 2011, China approved pilot tests of carbon trading in seven provinces and cities – Beijing, Chongqing, Shanghai, Shenzhen, Tianjin as well as Guangdong Province and Hubei Province, with different prices in each region. The pilot is intended to test the waters and provide valuable lessons for the design of a national system in the near future. Their successes or failures will therefore have far reaching implications for carbon market development in China in terms of trust in a national carbon trading market. Some of the pilot regions can start trading as early as 2013/2014. National trading is expected to start in 2017, latest in 2020.

India

Trading is set to begin in 2014 after a three-year rollout period. It is a mandatory energy efficiency trading scheme covering eight sectors responsible for 54 per cent of India's industrial energy consumption. India has pledged a 20 to 25 per cent reduction in emissions intensity from 2005 levels by 2020. Under the scheme, annual efficiency targets will be allocated to firms. Tradable energy-saving permits will be issued depending on the amount of energy saved during a target year.

Renewable Energy Certificates

Renewable Energy Certificates (occasionally referred to as or "green tags" [citation required]), are a largely unrelated form of market-based instruments that are used to achieve renewable energy targets, which may be environmentally motivated (like emissions reduction targets), but may also be motivated by other aims, such as energy security or industrial policy.

Carbon Market

Carbon emissions trading is emissions trading specifically for carbon dioxide (calculated in tonnes of carbon dioxide equivalent or tCO_2e) and currently makes up the bulk of emissions trading. It is one of the ways countries can meet their obligations under the Kyoto Protocol to reduce carbon emissions and thereby mitigate global warming.

Market Trend

Trading can be done directly between buyers and sellers, through several organised exchanges or through the many intermediaries active in the carbon market. The price of allowances is determined by supply and demand. As many as 40 million allowances have been traded per day. In 2012, 7.9 billion allowances were traded with a total value of €56 billion. Carbon emissions trading declined in 2013, and is expected to decline in 2014.

According to the World Bank's Carbon Finance Unit, 374 million metric tonnes of carbon dioxide equivalent (tCO_2e) were exchanged through projects in 2005, a 240% increase relative to 2004 (110 $mtCO_2e$) which was itself a 41% increase relative to 2003 (78 $mtCO_2e$).

Global carbon markets have shrunk in value by 60% since 2011, but are expected to rise again in 2014.

In terms of dollars, the World Bank has estimated that the size of the carbon market was 11 billion USD in 2005, 30 billion USD in 2006, and 64 billion in 2007.

The Marrakesh Accords of the Kyoto protocol defined the international trading mechanisms and registries needed to support trading between countries (*sources can buy or sell allowances on the open market. Because the total number of allowances is limited by the cap, emission reductions are assured.*). Allowance trading now occurs between European countries and Asian countries. However, while the USA as a nation did not ratify the Protocol, many of its states are developing cap-and-trade systems and considering ways to link them together, nationally and internationally, to find the lowest costs and improve liquidity of the market. However, these states also wish to preserve their individual integrity and unique features. For example, in contrast to other Kyoto-compliant systems, some states propose other types of greenhouse gas sources, different measurement methods, setting a maximum on the price of allowances, or restricting access to CDM projects. Creating instruments that are not fungible (exchangeable) could introduce instability and make pricing difficult. Various proposals for linking these systems across markets are being investigated, and this is being coordinated by the International Carbon Action Partnership (ICAP).

Business Reaction

In 2008, Barclays Capital predicted that the new carbon market would be worth $70 billion worldwide that year. The voluntary offset market, by comparison, is projected to grow to about $4bn by 2010.

23 multinational corporations came together in the G8 Climate Change Roundtable, a business group formed at the January 2005 World Economic Forum. The group included Ford, Toyota, British Airways, BP and Unilever. On June 9, 2005 the Group published a statement stating the need to act on climate change and stressing the importance of market-based solutions. It called on governments to establish "clear, transparent, and consistent price signals" through "creation of a long-term policy framework" that would include all major producers of greenhouse gases. By December 2007, this had grown to encompass 150 global businesses.

Business in the UK have come out strongly in support of emissions trading as a key tool to mitigate climate change, supported by NGOs. However, not all businesses favor a trading approach. On December 11, 2008, Rex Tillerson, the CEO of Exxonmobil, said a carbon tax is "a more direct, more transparent and more effective approach" than a cap-and-trade program, which he said, "inevitably introduces unnecessary cost and complexity". He also said that he hoped that the revenues from a carbon tax would be used to lower other taxes so as to be revenue neutral.

The International Air Transport Association, whose 230 member airlines comprise 93% of all international traffic, position is that trading should be based on "benchmarking", setting emissions levels based on industry averages, rather than "grandfathering", which would use individual companies' previous emissions levels to set their future permit allowances. They argue grandfathering "would penalise airlines that took early action to modernise their fleets, while a benchmarking approach, if designed properly, would reward more efficient operations".

Measuring, Reporting, Verification

Assuring compliance with an emissions trading scheme requires measuring, reporting and verification (MRV). Measurements are needed at each operator or installation. These measurements are reported to a regulator. For greenhouse gases, all trading countries maintain an inventory of emissions at national and installation level; in addition, trading groups within North America maintain inventories at the state level through The Climate Registry. For trading between regions, these inventories must be consistent, with equivalent units and measurement techniques.

In some industrial processes, emissions can be physically measured by inserting sensors and flow-meters in chimneys and stacks, but many types of activity rely on theoretical calculations instead of measurement. Depending on local legislation, measurements may require additional checks and verification by government or third party auditors, prior or post submission to the local regulator.

Enforcement

Another troublesome aspect of cap-and-trade is enforcement. Without effective MRV and enforcement, the value of allowances diminishes. Enforcement methods include fines and sanctions for polluters that have exceeded their allowances. Concerns include the cost of MRV and enforcement, and the risk that facilities may lie about actual emissions. The net effect of a corrupt reporting system or poorly managed or financed regulator may be a discount on emission costs, and a hidden increase in actual emissions.

According to Nordhaus, strict enforcement of the Kyoto Protocol is likely to be ob-served in those countries and industries covered by the EU ETS. Ellerman and Buchner (2007, p. 71) commented on the European Commission's (EC's) role in enforcing scarcity of permits with-in the EU ETS. This was done by the EC's reviewing the total number of permits that member states proposed that their industries be allocated. Based on institutional and enforcement considerations, Kruger et al. (2007, pp. 130–131) suggested that emissions trading within developing countries might not be a realistic goal in the near-term. Burniaux et al. (2008, p. 56) argued that due to the difficulty in enforcing international rules against sovereign states, development of the carbon market would require negotiation and consensus-building.

Criticism

Emissions trading has been criticised for a variety of reasons.

For example, in the popular science magazine *New Scientist*, Lohmann (2006) argued that trading pollution allowances should be avoided as a climate stabilization policy for several reasons. First, climate change requires more radical changes than previous pollution trading schemes such as the US SO_2 market. It requires reorganizing society and technology to "leave most remaining fossil fuels safely underground". Carbon trading schemes have tended to reward the heaviest polluters with 'windfall profits' when they are granted enough carbon credits to match historic production. Expensive long-term structural changes will not be made if there are cheaper sources of carbon credits which are often available from less developed countries, where they may be generated by local polluters at the expense of local communities.

Lohmann (2006b) supported conventional regulation, green taxes, and energy policies that are "justice-based" and "community-driven." According to Carbon Trade Watch (2009), carbon trading has had a "disastrous track record." The effectiveness of the EU ETS was criticized, and it was argued that the CDM had routinely favoured "environmentally ineffective and socially unjust projects."

Annie Leonard's 2009 documentary *The Story of Cap and Trade* criticized carbon emissions trading for the free permits to major polluters giving them unjust advantages, cheating in connection with carbon offsets, and as a distraction from the search for other solutions.

Offsets

Forest campaigner Jutta Kill (2006) of European environmental group FERN argued that offsets for emission reductions were not substitute for actual cuts in emissions. Kill stated that "[carbon] in trees is temporary: Trees can easily release carbon into the atmosphere through fire, disease, climatic changes, natural decay and timber harvesting."

Permit Supply Level

Regulatory agencies run the risk of issuing too many emission credits, which can result in a very low price on emission permits. This reduces the incentive that permit-liable firms have to cut back their emissions. On the other hand, issuing too few permits can result in an excessively high permit price. This an argument for a hybrid instrument having a price-floor, i.e., a minimum permit price, and a price-ceiling, i.e., a limit on the permit price. However, a price-ceiling removes the certainty of a particular quantity limit of emissions.

Permit Allocation Versus Auctioning

If polluters receive emission permits for free ("grandfathering"), this may be a reason for them not to cut their emissions because if they do they will receive fewer permits in the future (IMF, 2008, pp. 25–26).

This perverse incentive can be alleviated if permits are auctioned, i.e., sold to polluters, rather than giving them the permits for free (Hepburn, 2006, pp. 236–237). Auctioning is a method for distributing emission allowances in a cap-and-trade system whereby allowances are sold to the highest bidder. Revenues from auctioning go to the government and can be used for development of sustainable technology or to cut distortionary taxes, thus improving the efficiency of the overall cap policy (Fisher *et al..*, 1996, p. 417).

On the other hand, allocating permits can be used as a measure to protect domestic firms who are internationally exposed to competition (p. 237). This happens when domestic firms compete against other firms that are not subject to the same regulation. This argument in favor of allocation of permits has been used in the EU ETS, where industries that have been judged to be internationally exposed, e.g., cement and steel production, have been given permits for free (4CMR, 2008).

This method of distribution may be combined with other forms of allowance distribution.

Distributional Effects

The US Congressional Budget Office (CBO, 2009) examined the potential effects of the American Clean Energy and Security Act on US households. This act relies heavily on the free allocation of permits. The Bill was found to protect low-income consumers, but it was recommended that the Bill be made more efficient by reducing welfare provisions for corporations, and more resources be made available for consumer relief.

Linking

Distinct cap-and-trade systems can be linked together through the mutual or unilateral recognition of emissions allowances for compliance. Linking systems creates a larger carbon market, which can reduce overall compliance costs, increase market liquidity and generate a more stable carbon market. Linking systems can also be politically symbolic as it shows willingness to undertake a common effort to reduce GHG emissions. Some scholars have argued that linking may provide a starting point for developing a new, bottom-up international climate policy architecture, whereby multiple unique systems successively link their various systems. In 2014, the U.S. state of California and the Canadian province of Québec successfully linked their systems. In 2015, the provinces of Ontario and Manitoba agreed to join the linked system between Quebec and California.

The International Carbon Action Partnership brings together regional, national and sub-national governments and public authorities from around the world to discuss important issues in the design of emissions trading schemes (ETS) and the way forward to a global carbon market. 30 national and subnational jurisdictions have joined ICAP as members since its establishment in 2007.

Greenhouse Gas Removal

Greenhouse gas removal projects are a type of climate engineering that seek to remove greenhouse gases from the atmosphere, and thus they tackle the root cause of global warming. These techniques either directly remove greenhouse gases, or alternatively seek to influence natural processes to remove greenhouse gases indirectly. The discipline overlaps with carbon capture and storage and carbon sequestration, and some projects listed may not be considered to be geoengineering by all commentators, instead being described as mitigation.

Carbon Sequestration

A wide range of techniques for carbon sequestration exist. These range from ideas to remove CO_2 from the atmosphere (carbon dioxide air capture), flue gases (carbon capture and storage) and by preventing carbon in biomass from re-entering the atmosphere, such as with Bio-energy with carbon capture and storage (BECCS).

Chlorofluorocarbon Photochemistry

Atmospheric chlorofluorocarbon (CFC) removal is an idea which suggests using lasers to break up CFCs, an important family of greenhouse gases, in the atmosphere.

Methane Removal

Methane potentially poses major challenges for remediation. It is around 20 times as powerful a greenhouse gas as CO_2. Large quantities may be outgassed from permafrost and clathrates as a result of global warming, notably in the Arctic.

There are existing climate engineering proposals. Methane is removed by several natural processes, which can be enhanced.

- Chemical decomposition — reaction with hydroxyl radicals produced from photochemical decomposition of ozone in the stratosphere.

- Biological decomposition — by methanotrophs in soils and water.

References

- Sullivan, Arthur, and Steven M. Sheffrin. Economics: Principles in action. Upper Saddle River, New Jersey, 2003. ISBN 0-13-063085-3

- Rosen and Gayer, Harvey S. and Ted (2008). Public Finance. New York: McGraw-Hill Irwin. pp. 90–94. ISBN 978-0-07-351128-3.

- Burney, Nelson E. (2010). Carbon Tax and Cap-and-trade Tools : Market-based Approaches for Controlling Greenhouse Gases. New York: Nova Science Publishers, Inc. ISBN 9781608761371.

- Dryzek, John S.; Norgaard, Richard B.; Schlosberg, David (2011). The Oxford Handbook of Climate Change and Society. Oxford University Press. p. 154. ISBN 9780199683420.

- Tiwari, Gopal Nath; Agrawal, Basant (2010). Building integrated photovoltaic thermal systems : for sustainable developments. Cambridge: Royal Society of Chemistry. ISBN 1-84973-090-3.

- Stix, T.H. (7–9 Jun 1993). "Removal of chlorofluorocarbons from the troposphere". 1993 IEEE International Conference on Plasma Science. Vancouver, BC, Canada: IEEE. p. 135. doi:10.1109/PLASMA.1993.593398. ISBN 0-7803-1360-7.

- "2050 Ett koldioxidneutralt Sverige". naturvardsverket.se. 2013-02-30. Retrieved 2016-01-30. Check date Mayer, Jane (August 30, 2010). "Covert Operations: The billionaire brothers who are waging a war against Obama". The New Yorker. Retrieved March 20, 2015.

- "International Carbon Action Partnership (ICAP)". International Carbon Action Partnership. Retrieved 25 September 2015.

- "Carbon Place . EU - Market with carbon credits - EUA, CER, ERU, VER, AAU-S, AAU-G". Retrieved 25 September 2015.

- "Key points: Update Paper 6: Carbon pricing and reducing Australia's emissions". Garnaut Climate Change Review. 17 March 2011. Retrieved 16 July 2013.

- Ranson, M., & Stavins, R., 2013: Linkage of Greenhouse Gas Emissions Trading Systems - Learning from Experience. Discussion Paper Resources For The Future, No. 42.

Role of Greenhouse Gas Emission in Global Warming and Climate Change

Greenhouse gases contribute greatly to the recent change in climate and there is an arising need to assess it in terms of global warming and climate change. National and international policies and protocols based on greenhouse gas emissions depend on these projections for policy-making and trade.

Attribution of Recent Climate Change

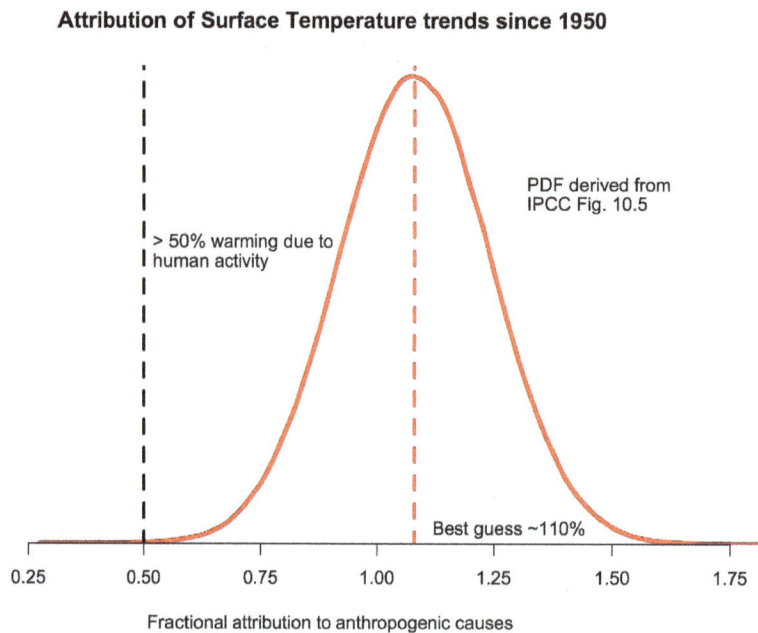

Attribution of Surface Temperature trends since 1950

PDF derived from IPCC Fig. 10.5

> 50% warming due to human activity

Best guess ~110%

0.25   0.50   0.75   1.00   1.25   1.50   1.75

Fractional attribution to anthropogenic causes

PDF of fraction of surface temperature trends since 1950 attributable to human activity, based on IPCC AR5 10.5

Global annual average temperature; year-to-year fluctuations are due to natural processes, such as the effects of El Niños, La Niñas, and the eruption of large volcanoes.

This image shows three examples of internal climate variability measured between 1950 and 2012: the El Niño–Southern oscillation, the Arctic oscillation, and the North Atlantic oscillation.

Attribution of recent climate change is the effort to scientifically ascertain mechanisms responsible for recent climate changes on Earth, commonly known as 'global warming'. The effort has focused on changes observed during the period of instrumental temperature record, when records are most reliable; particularly in the last 50 years, when human activity has grown fastest and observations of the troposphere have become available. The dominant mechanisms (to which the IPCC attributes climate change) are anthropogenic, i.e., the result of human activity. They are:

- increasing atmospheric concentrations of greenhouse gases

- global changes to land surface, such as deforestation

- increasing atmospheric concentrations of aerosols.

There are also natural mechanisms for variation including climate oscillations, changes in solar activity, and volcanic activity.

According to the Intergovernmental Panel on Climate Change (IPCC), it is "extremely likely" that human influence was the dominant cause of global warming between 1951 and 2010. The IPCC defines "extremely likely" as indicating a probability of 95 to 100%, based on an expert assessment of all the available evidence.

Multiple lines of evidence support attribution of recent climate change to human activities:

- A basic physical understanding of the climate system: greenhouse gas concentrations have increased and their warming properties are well-established.

- Historical estimates of past climate changes suggest that the recent changes in global surface temperature are unusual.

- Computer-based climate models are unable to replicate the observed warming unless human greenhouse gas emissions are included.

- Natural forces alone (such as solar and volcanic activity) cannot explain the observed warming.

The IPCC's attribution of recent global warming to human activities is a view shared by most scientists, and is also supported by 196 other scientific organizations worldwide.

Background

This section introduces some concepts in climate science that are used in the following sections:

Factors affecting Earth's climate can be broken down into feedbacks and forcings. A forcing is something that is imposed externally on the climate system. External forcings include natural phenomena such as volcanic eruptions and variations in the sun's output. Human activities can also impose forcings, for example, through changing the composition of the atmosphere.

Radiative forcing is a measure of how various factors alter the energy balance of the Earth's atmosphere. A positive radiative forcing will tend to increase the energy of the Earth-atmosphere system, leading to a warming of the system. Between the start of the Industrial Revolution in 1750, and the year 2005, the increase in the atmospheric concentration of carbon dioxide (chemical formula: CO_2) led to a positive radiative forcing, averaged over the Earth's surface area, of about 1.66 watts per square metre (abbreviated W m^{-2}).

Climate feedbacks can either amplify or dampen the response of the climate to a given forcing.

There are many feedback mechanisms in the climate system that can either amplify (a positive feedback) or diminish (a negative feedback) the effects of a change in climate forcing.

Aspects of the climate system will show variation in response to changes in forcings. In the absence of forcings imposed on it, the climate system will still show internal variability. This internal variability is a result of complex interactions between components of the climate system, such as the coupling between the atmosphere and ocean. An example of internal variability is the El Niño-Southern Oscillation.

Detection Vs. Attribution

In detection and attribution, the natural factors considered usually include changes in the Sun's output and volcanic eruptions, as well as natural modes of variability such as El Niño and La Niña. Human factors include the emissions of heat-trapping "greenhouse" gases and particulates as well as clearing of forests and other land-use changes. Figure source: NOAA NCDC.

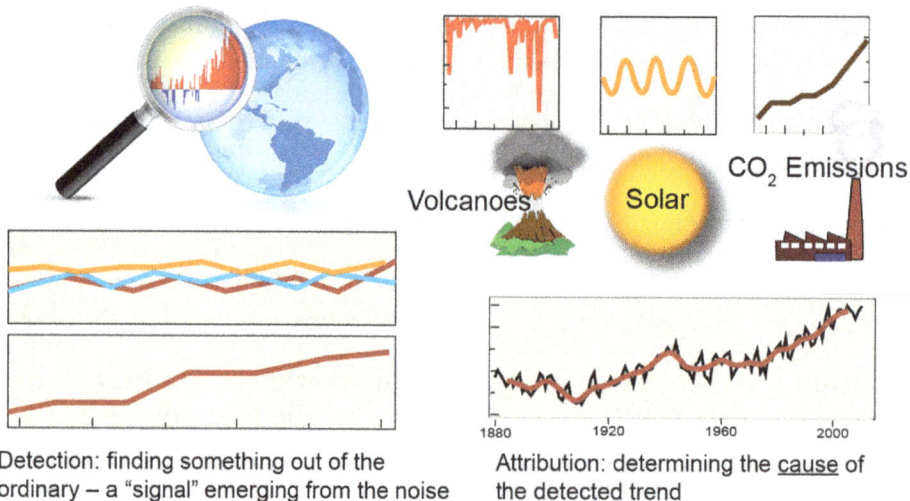

Detection and Attribution as Forensics

Detection: finding something out of the ordinary – a "signal" emerging from the noise

Attribution: determining the cause of the detected trend

In detection and attribution, the natural factors considered usually include changes in the Sun's output and volcanic eruptions, as well as natural modes of variability such as El Niño and La Niña. Human factors include the emissions of heat-trapping "greenhouse" gases and particulates as well as clearing of forests and other land-use changes. Figure source: NOAA NCDC.

Detection and attribution of climate signals, as well as its common-sense meaning, has a more precise definition within the climate change literature, as expressed by the IPCC. Detection of a climate signal does not always imply significant attribution. The IPCC's Fourth Assessment Report says "it is *extremely likely* that human activities have exerted a substantial net warming influence on climate since 1750," where "extremely likely" indicates a probability greater than 95%. *Detection* of a signal requires demonstrating that an observed change is statistically significantly different from that which can be explained by natural internal variability.

Attribution requires demonstrating that a signal is:

- unlikely to be due entirely to internal variability;

- consistent with the estimated responses to the given combination of anthropogenic and natural forcing

- not consistent with alternative, physically plausible explanations of recent climate change that exclude important elements of the given combination of forcings.

Key Attributions

Greenhouse Gases

Carbon dioxide is the primary greenhouse gas that is contributing to recent climate change. CO2 is absorbed and emitted naturally as part of the carbon cycle, through animal and plant respiration, volcanic eruptions, and ocean-atmosphere exchange. Human activities, such as the burning of fossil fuels and changes in land use, release large amounts of carbon to the atmosphere, causing CO2 concentrations in the atmosphere to rise.

The high-accuracy measurements of atmospheric CO_2 concentration, initiated by Charles David Keeling in 1958, constitute the master time series documenting the changing composition of the atmosphere. These data have iconic status in climate change science as evidence of the effect of human activities on the chemical composition of the global atmosphere.

Along with CO_2, methane and nitrous oxide are also major forcing contributors to the greenhouse effect. The Kyoto Protocol lists these together with hydrofluorocarbons (HFCs), perfluorocarbons (PFCs), and sulphur hexafluoride (SF_6), which are entirely artificial (i.e. anthropogenic) gases, which also contribute to radiative forcing in the atmosphere. The chart at right attributes anthropogenic greenhouse gas emissions to eight main economic sectors, of which the largest contributors are power stations (many of which burn coal or other fossil fuels), industrial processes, transportation fuels (generally fossil fuels), and agricultural by-products (mainly methane from enteric fermentation and nitrous oxide from fertilizer use).

Water Vapor

Water vapor is the most abundant greenhouse gas and also the most important in terms of its contribution to the natural greenhouse effect, despite having a short atmospheric lifetime (about 10 days). Some human activities can influence local water vapor levels. However, on a global scale, the concentration of water vapor is controlled by temperature, which influences overall rates of evaporation and precipitation. Therefore, the global concentration of water vapor is not substantially affected by direct human emissions.

Land Use

Climate change is attributed to land use for two main reasons. Between 1750 and 2007, about two-thirds of anthropogenic CO_2 emissions were produced from burning fossil fuels, and about one-third of emissions from changes in land use, primarily deforestation. Deforestation both reduces the amount of carbon dioxide absorbed by deforested regions and releases greenhouse gases directly, together with aerosols, through biomass burning that frequently accompanies it.

A second reason that climate change has been attributed to land use is that the terrestrial albedo is often altered by use, which leads to radiative forcing. This effect is more significant locally than globally.

Livestock and Land Use

Worldwide, livestock production occupies 70% of all land used for agriculture, or 30% of the ice-free land surface of the Earth. More than 18% of anthropogenic greenhouse gas emissions are attributed to livestock and livestock-related activities such as deforestation and increasingly fuel-intensive farming practices. Specific attributions to the livestock sector include:

- 9% of global anthropogenic carbon dioxide emissions

- 35–40% of global anthropogenic methane emissions (chiefly due to enteric fermentation and manure)

- 64% of global anthropogenic nitrous oxide emissions, chiefly due to fertilizer use.

Aerosols

With virtual certainty, scientific consensus has attributed various forms of climate change, chiefly cooling effects, to aerosols, which are small particles or droplets suspended in the atmosphere. Key sources to which anthropogenic aerosols are attributed include:

- biomass burning such as slash and burn deforestation. Aerosols produced are primarily black carbon.

- industrial air pollution, which produces soot and airborne sulfates, nitrates, and ammonium

- dust produced by land use effects such as desertification

Attribution of 20th Century Climate Change

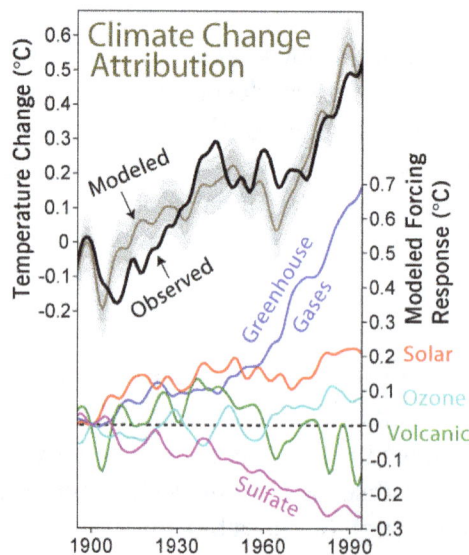

One global climate model's reconstruction of temperature change during the 20th century as the result of five studied forcing factors and the amount of temperature change attributed to each.

Over the past 150 years human activities have released increasing quantities of greenhouse gases into the atmosphere. This has led to increases in mean global temperature, or global warming. Other human effects are relevant—for example, sulphate aerosols are believed to have a cooling effect. Natural factors also contribute. According to the historical temperature record of the last century, the Earth's near-surface air temperature has risen around 0.74 ± 0.18 °Celsius (1.3 ± 0.32 °Fahrenheit).

A historically important question in climate change research has regarded the relative importance of human activity and non-anthropogenic causes during the period of instrumental record. In the 1995 Second Assessment Report (SAR), the IPCC made the widely quoted statement that "The balance of evidence suggests a discernible human influence on global climate". The phrase "balance of evidence" suggested the (English) common-law standard of proof required in civil as opposed to criminal courts: not as high as "beyond reasonable doubt". In 2001 the Third Assessment Report (TAR) refined this, saying "There is new and stronger evidence that most of the warming observed over the last 50 years is attributable to human activities". The 2007 Fourth Assessment Report (AR4) strengthened this finding:

- "Anthropogenic warming of the climate system is widespread and can be detected in temperature observations taken at the surface, in the free atmosphere and in the oceans. Evidence of the effect of external influences, both anthropogenic and natural, on the climate system has continued to accumulate since the TAR."

Other findings of the IPCC Fourth Assessment Report include:

- "It is *extremely unlikely* (<5%) that the global pattern of warming during the past half century can be explained without external forcing (i.e., it is inconsistent with being the result of internal variability), and *very unlikely* that it is due to known natural external causes alone. The warming occurred in both the ocean and the atmosphere and took place at a time when natural external forcing factors would likely have produced cooling."

- "From new estimates of the combined anthropogenic forcing due to greenhouse gases, aerosols, and land surface changes, it is *extremely likely* (>95%) that human activities have exerted a substantial net warming influence on climate since 1750."

- "It is *virtually certain* that anthropogenic aerosols produce a net negative radiative forcing (cooling influence) with a greater magnitude in the Northern Hemisphere than in the Southern Hemisphere."

Over the past five decades there has been a global warming of approximately 0.65 °C (1.17 °F) at the Earth's surface. Among the possible factors that could produce changes in global mean temperature are internal variability of the climate system, external forcing, an increase in concentration of greenhouse gases, or any combination of these. Current studies indicate that the increase in greenhouse gases, most notably CO_2, is mostly responsible for the observed warming. Evidence for this conclusion includes:

- Estimates of internal variability from climate models, and reconstructions of past temperatures, indicate that the warming is unlikely to be entirely natural.

- Climate models forced by natural factors *and* increased greenhouse gases and aerosols reproduce the observed global temperature changes; those forced by natural factors alone do not.

- "Fingerprint" methods indicate that the pattern of change is closer to that expected from greenhouse gas-forced change than from natural change.

- The plateau in warming from the 1940s to 1960s can be attributed largely to sulphate aerosol cooling.

Details on Attribution

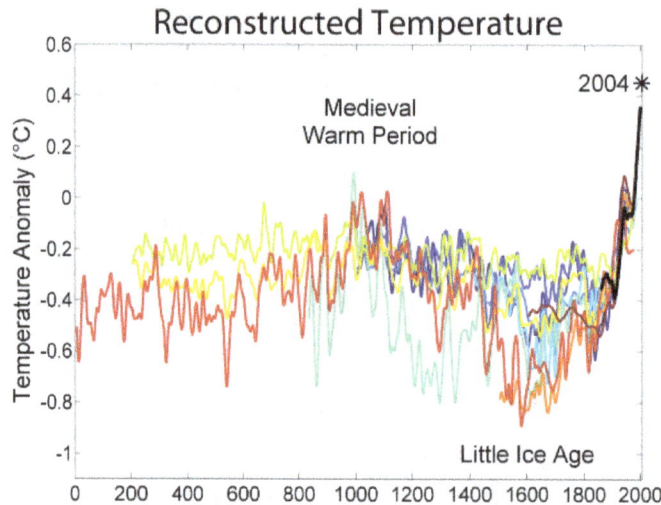

Reconstructed Temperature

For Northern Hemisphere temperature, recent decades appear to be the warmest since at least about 1000AD, and the warming since the late 19th century is unprecedented over the last 1000 years. Older data are insufficient to provide reliable hemispheric temperature estimates.

Recent scientific assessments find that most of the warming of the Earth's surface over the past 50 years has been caused by human activities. This conclusion rests on multiple lines of evidence. Like the warming "signal" that has gradually emerged from the "noise" of natural climate variability, the scientific evidence for a human influence on global climate has accumulated over the past several decades, from many hundreds of studies. No single study is a "smoking gun." Nor has any single study or combination of studies undermined the large body of evidence supporting the conclusion that human activity is the primary driver of recent warming.

The first line of evidence is based on a physical understanding of how greenhouse gases trap heat, how the climate system responds to increases in greenhouse gases, and how other human and natural factors influence climate. The second line of evidence is from indirect estimates of climate changes over the last 1,000 to 2,000 years. These records are obtained from living things and their remains (like tree rings and corals) and from physical quantities (like the ratio between lighter and heavier isotopes of oxygen in ice cores), which change in measurable ways as climate changes. The lesson from these data is that global surface temperatures over the last several decades are clearly unusual, in that they were higher than at any time during at least the past 400 years. For the Northern Hemisphere, the recent temperature rise is clearly unusual in at least the last 1,000 years.

The third line of evidence is based on the broad, qualitative consistency between observed changes in climate and the computer model simulations of how climate would be expected to change in response to human activities. For example, when climate models are run with historical increases in greenhouse gases, they show gradual warming of the Earth and ocean surface, increases in ocean heat content and the temperature of the lower atmosphere, a rise in global sea level, retreat of sea ice and snow cover, cooling of the stratosphere, an increase in the amount of atmospheric water vapor, and changes in large-scale precipitation and pressure patterns. These and other aspects of modelled climate change are in agreement with observations.

"Fingerprint" Studies

Reconstructions of global temperature that include greenhouse gas increases and other human influences (red line, based on many models) closely match measured temperatures (dashed line). Those that only include natural influences (blue line, based on many models) show a slight cooling, which has not occurred. The ability of models to generate reasonable histories of global temperature is verified by their response to four 20th-century volcanic eruptions: each eruption caused brief cooling that appeared in observed as well as modeled records.

Finally, there is extensive statistical evidence from so-called "fingerprint" studies. Each factor that affects climate produces a unique pattern of climate response, much as each person has a unique fingerprint. Fingerprint studies exploit these unique signatures, and allow detailed comparisons of modelled and observed climate change patterns. Scientists rely on such studies to attribute observed changes in climate to a particular cause or set of causes. In the real world, the climate changes that have occurred since the start of the Industrial Revolution are due to a complex mixture of human and natural causes. The importance of each individual influence in this mixture changes over time. Of course, there are not multiple Earths, which would allow an experimenter to change one factor at a time on each Earth, thus helping to isolate different fingerprints. Therefore, climate models are used to study how individual factors affect climate. For example, a single factor (like greenhouse gases) or a set of factors can be varied, and the response of the modelled climate system to these individual or combined changes can thus be studied.

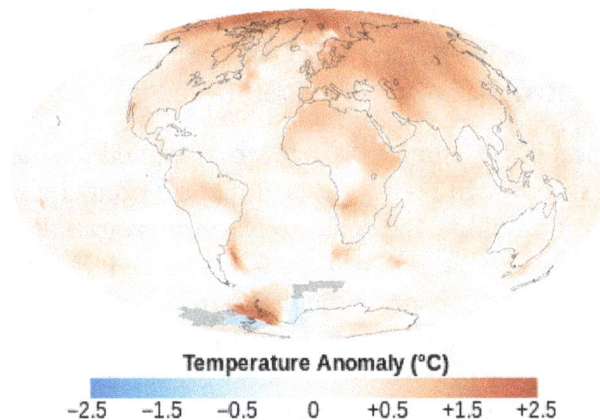

Temperature Anomaly (°C)

-2.5   -1.5   -0.5   0   +0.5   +1.5   +2.5

Two fingerprints of human activities on the climate are that land areas will warm more than the oceans, and that high latitudes will warm more than low latitudes. These projections have been confirmed by observations (shown above).

For example, when climate model simulations of the last century include all of the major influences on climate, both human-induced and natural, they can reproduce many important features of observed climate change patterns. When human influences are removed from the model experiments, results suggest that the surface of the Earth would actually have cooled slightly over the last 50 years. The clear message from fingerprint studies is that the observed warming over the last half-century cannot be explained by natural factors, and is instead caused primarily by human factors.

Another fingerprint of human effects on climate has been identified by looking at a slice through the layers of the atmosphere, and studying the pattern of temperature changes from the surface up through the stratosphere. The earliest fingerprint work focused on changes in surface and atmospheric temperature. Scientists then applied fingerprint methods to a whole range of climate variables, identifying human-caused climate signals in the heat content of the oceans, the height of the tropopause (the boundary between the troposphere and stratosphere, which has shifted upward by hundreds of feet in recent decades), the geographical patterns of precipitation, drought, surface pressure, and the runoff from major river basins.

Studies published after the appearance of the IPCC Fourth Assessment Report in 2007 have also found human fingerprints in the increased levels of atmospheric moisture (both close to the surface and over the full extent of the atmosphere), in the decline of Arctic sea ice extent, and in the patterns of changes in Arctic and Antarctic surface temperatures.

The message from this entire body of work is that the climate system is telling a consistent story of increasingly dominant human influence – the changes in temperature, ice extent, moisture, and circulation patterns fit together in a physically consistent way, like pieces in a complex puzzle.

Increasingly, this type of fingerprint work is shifting its emphasis. As noted, clear and compelling scientific evidence supports the case for a pronounced human influence on global climate. Much of the recent attention is now on climate changes at continental and regional scales, and on variables that can have large impacts on societies. For example, scientists have established causal links between human activities and the changes in snowpack, maximum and minimum (diurnal) temperature, and the seasonal timing of runoff over mountainous regions of the western United States. Human activity is likely to have made a substantial contribution to ocean surface temperature changes in hurricane formation regions. Researchers are also looking beyond the physical climate system, and are beginning to tie changes in the distribution and seasonal behaviour of plant and animal species to human-caused changes in temperature and precipitation.

For over a decade, one aspect of the climate change story seemed to show a significant difference between models and observations. In the tropics, all models predicted that with a rise in greenhouse gases, the troposphere would be expected to warm more rapidly than the surface. Observations from weather balloons, satellites, and surface thermometers seemed to show the opposite behaviour (more rapid warming of the surface than the troposphere). This issue was a stumbling block in understanding the causes of climate change. It is now largely resolved. Research showed that there were large uncertainties in the satellite and weather balloon data. When uncertainties in models and observations are properly accounted for, newer observational data sets (with better treatment of known problems) are in agreement with climate model results.

This set of graphs shows the estimated contribution of various natural and human factors to changes in global mean temperature between 1889–2006. Estimated contributions are based on multivariate analysis rather than model simulations. The graphs show that human influence on climate has eclipsed the magnitude of natural temperature changes over the past 120 years. Natural influences on temperature—El Niño, solar variability, and volcanic aerosols—have varied approximately plus and minus 0.2 °C (0.4 °F), (averaging to about zero), while human influences have contributed roughly 0.8 °C (1 °F) of warming since 1889.

This does not mean, however, that all remaining differences between models and observations have been resolved. The observed changes in some climate variables, such as Arctic sea ice, some aspects of precipitation, and patterns of surface pressure, appear to be proceeding much more rapidly than models have projected. The reasons for these differences are not well understood. Nevertheless, the bottom-line conclusion from climate fingerprinting is that most of the observed changes studied to date are consistent with each other, and are also consistent with our scientific understanding of how the climate system would be expected to respond to the increase in heat-trapping gases resulting from human activities.

Extreme Weather Events

SHIFTING DISTRIBUTION OF SUMMER TEMPERATURE ANOMALIES

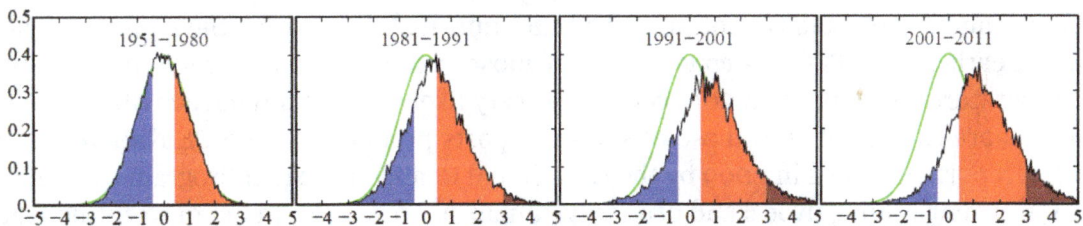

Credit: James Hansen, NASA Goddard Institute for Space Studies

Frequency of occurrence (vertical axis) of local June–July–August temperature anomalies (relative to 1951–1980 mean) for Northern Hemisphere land in units of local standard deviation (horizontal axis). According to Hansen *et al.* (2012), the distribution of anomalies has shifted to the right as a consequence of global warming, meaning that unusually hot summers have become more common. This is analogous to the rolling of a dice: cool summers now cover only half of one side of a six-sided die, white covers one side, red covers four sides, and an extremely hot (redbrown) anomaly covers half of one side.

One of the subjects discussed in the literature is whether or not extreme weather events can be attributed to human activities. Seneviratne *et al.* (2012) stated that attributing individual extreme weather events to human activities was challenging. They were, however, more confident over attributing changes in long-term trends of extreme weather. For example, Seneviratne *et al.* (2012) concluded that human activities had likely led to a warming of extreme daily minimum and maximum temperatures at the global scale.

Another way of viewing the problem is to consider the effects of human-induced climate change on the probability of future extreme weather events. Stott *et al.* (2003), for example, considered whether or not human activities had increased the risk of severe heat waves in Europe, like the one experienced in 2003. Their conclusion was that human activities had very likely more than doubled the risk of heat waves of this magnitude.

An analogy can be made between an athlete on steroids and human-induced climate change. In the same way that an athlete's performance may increase from using steroids, human-induced climate change increases the risk of some extreme weather events.

Hansen *et al.* (2012) suggested that human activities have greatly increased the risk of summertime heat waves. According to their analysis, the land area of the Earth affected by very hot summer temperature anomalies has greatly increased over time (refer to graphs on the left). In the base period 1951-1980, these anomalies covered a few tenths of 1% of the global land area. In recent years, this has increased to around 10% of the global land area. With high confidence, Hansen *et al.* (2012) attributed the 2010 Moscow and 2011 Texas heat waves to human-induced global warming.

An earlier study by Dole *et al.* (2011) concluded that the 2010 Moscow heatwave was mostly due to natural weather variability. While not directly citing Dole *et al.* (2011), Hansen *et al.* (2012) rejected this type of explanation. Hansen *et al.* (2012) stated that a combination of natural weather variability and human-induced global warming was responsible for the Moscow and Texas heat waves.

Scientific Literature and Opinion

There are a number of examples of published and informal support for the consensus view. As mentioned earlier, the IPCC has concluded that most of the observed increase in globally averaged temperatures since the mid-20th century is "very likely" due to human activities. The IPCC's conclusions are consistent with those of several reports produced by the US National Research Council. A report published in 2009 by the U.S. Global Change Research Program concluded that "[global] warming is unequivocal and primarily human-induced." A number of scientific organizations have issued statements that support the consensus view. Two examples include:

- a joint statement made in 2005 by the national science academies of the G8, and Brazil, China and India;

- a joint statement made in 2008 by the Network of African Science Academies.

Detection and Attribution Studies

The IPCC Fourth Assessment Report (2007), concluded that attribution was possible for a number of observed changes in the climate. However, attribution was found to be more difficult when assessing changes over smaller regions (less than continental scale) and over short time periods (less than 50 years). Over larger regions, averaging reduces natural variability of the climate, making detection and attribution easier.

- In 1996, in a paper in *Nature* titled "A search for human influences on the thermal structure of the atmosphere", Benjamin D. Santer et al. wrote: "The observed spatial patterns of temperature change in the free atmosphere from 1963 to 1987 are similar to those predicted by state-of-the-art climate models incorporating various combinations of changes in carbon dioxide, anthropogenic sulphate aerosol and stratospheric ozone concentrations. The degree of pattern similarity between models and observations increases through this period. It is likely that this trend is partially due to human activities, although many uncertainties remain, particularly relating to estimates of natural variability."

- A 2002 paper in the *Journal of Geophysical Research* says "Our analysis suggests that the early twentieth century warming can best be explained by a combination of warming due to increases in greenhouse gases and natural forcing, some cooling due to other anthropogenic forcings, and a substantial, but not implausible, contribution from internal variability. In the second half of the century we find that the warming is largely caused by changes in greenhouse gases, with changes in sulphates and, perhaps, volcanic aerosol offsetting approximately one third of the warming."

- A 2005 review of detection and attribution studies by the International Ad Hoc Detection and Attribution Group found that "natural drivers such as solar variability and volcanic activity are at most partially responsible for the large-scale temperature changes observed over the past century, and that a large fraction of the warming over the last 50 yr can be attributed to greenhouse gas increases. Thus, the recent research supports and strengthens the IPCC Third Assessment Report conclusion that 'most of the global warming over the past 50 years is likely due to the increase in greenhouse gases.'"

- Barnett and colleagues (2005) say that the observed warming of the oceans "cannot be explained by natural internal climate variability or solar and volcanic forcing, but is well simulated by two anthropogenically forced climate models," concluding that "it is of human origin, a conclusion robust to observational sampling and model differences".

- Two papers in the journal *Science* in August 2005 resolve the problem, evident at the time of the TAR, of tropospheric temperature trends. The UAH version of the record contained errors, and there is evidence of spurious cooling trends in the radiosonde record, particularly in the tropics.

- Multiple independent reconstructions of the temperature record of the past 1000 years confirm that the late 20th century is probably the warmest period in that time.

Reviews of Scientific Opinion

- An essay in *Science* surveyed 928 abstracts related to climate change, and concluded that most journal reports accepted the consensus. This is discussed further in scientific opinion on climate change.

- A 2010 paper in the Proceedings of the National Academy of Sciences found that among a pool of roughly 1,000 researchers who work directly on climate issues and publish the most frequently on the subject, 97% agree that anthropogenic climate change is happening.

- A 2011 paper from George Mason University published in the *International Journal of Public Opinion Research*, "The Structure of Scientific Opinion on Climate Change," collected the opinions of scientists in the earth, space, atmospheric, oceanic or hydrological sciences. The 489 survey respondents—representing nearly half of all those eligible according to the survey's specific standards – work in academia, government, and industry, and are members of prominent professional organizations. The study found that 97% of the 489 scientists surveyed agreed that global temperatures have risen over the past century.

Moreover, 84% agreed that "human-induced greenhouse warming" is now occurring." Only 5% disagreed with the idea that human activity is a significant cause of global warming.

As described above, a small minority of scientists do disagree with the consensus. For example, Willie Soon and Richard Lindzen say that there is insufficient proof for anthropogenic attribution. Generally this position requires new physical mechanisms to explain the observed warming.

Solar Activity

Solar radiation at the top of our atmosphere, and global temperature

Modelled simulation of the effect of various factors (including GHGs, Solar irradiance) singly and in combination, showing in particular that solar activity produces a small and nearly uniform warming, unlike what is observed.

Solar sunspot maximum occurs when the magnetic field of the sun collapses and reverse as part of its average 11 year solar cycle (22 years for complete North to North restoration).

The role of the sun in recent climate change has been looked at by climate scientists. Since 1978, output from the Sun has been measured by satellites significantly more accurately than was previously possible from the surface. These measurements indicate that the Sun's total solar irradiance has not increased since 1978, so the warming during the past 30 years cannot be directly attributed to an increase in total solar energy reaching the Earth. In the three decades since 1978, the combination of solar and volcanic activity probably had a slight cooling influence on the climate.

Climate models have been used to examine the role of the sun in recent climate change. Models are unable to reproduce the rapid warming observed in recent decades when they only take into account variations in total solar irradiance and volcanic activity. Models are, however, able to simulate the observed 20th century changes in temperature when they include all of the most important external forcings, including human influences and natural forcings. As has already been stated, Hegerl *et al.* (2007) concluded that greenhouse gas forcing had "very likely" caused most of the observed global warming since the mid-20th century. In making this conclusion, Hegerl *et al.* (2007) allowed for the possibility that climate models had been underestimated the effect of solar forcing.

The role of solar activity in climate change has also been calculated over longer time periods using "proxy" datasets, such as tree rings. Models indicate that solar and volcanic forcings can explain periods of relative warmth and cold between A.D. 1000 and 1900, but human-induced forcings are needed to reproduce the late-20th century warming.

Another line of evidence against the sun having caused recent climate change comes from looking at how temperatures at different levels in the Earth's atmosphere have changed. Models and observations show that greenhouse gas results in warming of the lower atmosphere at the surface (called the troposphere) but cooling of the upper atmosphere (called the stratosphere). Depletion of the ozone layer by chemical refrigerants has also resulted in a cooling effect in the stratosphere. If the sun was responsible for observed warming, warming of the troposphere at the surface and warming at the top of the stratosphere would be expected as increase solar activity would replenish ozone and oxides of nitrogen. The stratosphere has a reverse temperature gradient than the troposphere so as the temperature of the troposphere cools with altitude, the stratosphere rises with altitude. Hadley cells are the mechanism by which equatorial generated ozone in the tropics (highest area of UV irradiance in the stratosphere) is moved poleward. Global climate models suggest that climate change may widen the Hadley cells and push the jetstream northward thereby expanding the tropics region and resulting in warmer, dryer conditions in those areas overall.

Non-consensus Views

Habibullo Abdussamatov (2004), head of space research at St. Petersburg's Pulkovo Astronomical Observatory in Russia, has argued that the sun is responsible for recently observed climate change. Journalists for news sources canada.com (Solomon, 2007b), National Geographic News (Ravillious, 2007), and LiveScience (Than, 2007) reported on the story of warming on Mars. In these articles, Abdussamatov was quoted. He stated that warming on Mars was evidence that global warming on Earth was being caused by changes in the sun.

Ravillious (2007) quoted two scientists who disagreed with Abdussamatov: Amato Evan, a climate scientist at the University of Wisconsin-Madison, in the US, and Colin Wilson, a planetary physicist at Oxford University in the UK. According to Wilson, "Wobbles in the orbit of Mars are the main cause of its climate change in the current era". Than (2007) quoted Charles Long, a climate physicist at Pacific Northwest National Laboratories in the US, who disagreed with Abdussamatov.

Than (2007) pointed to the view of Benny Peiser, a social anthropologist at Liverpool John Moores University in the UK. In his newsletter, Peiser had cited a blog that had commented on warming observed on several planetary bodies in the Solar system. These included Neptune's moon Triton, Jupiter, Pluto and Mars. In an e-mail interview with Than (2007), Peiser stated that:

"I think it is an intriguing coincidence that warming trends have been observed on a number of very diverse planetary bodies in our solar system, (...) Perhaps this is just a fluke."

Than (2007) provided alternative explanations of why warming had occurred on Triton, Pluto, Jupiter and Mars.

The US Environmental Protection Agency (US EPA, 2009) responded to public comments on climate change attribution. A number of commenters had argued that recent climate change could be attributed to changes in solar irradiance. According to the US EPA (2009), this attribution was not supported by the bulk of the scientific literature. Citing the work of the IPCC (2007), the US EPA pointed to the low contribution of solar irradiance to radiative forcing since the start of the Industrial Revolution in 1750. Over this time period (1750 to 2005), the estimated contribution of solar irradiance to radiative forcing was 5% the value of the combined radiative forcing due to increases in the atmospheric concentrations of carbon dioxide, methane and nitrous oxide.

Effect of Cosmic Rays

Henrik Svensmark has suggested that the magnetic activity of the sun deflects cosmic rays, and that this may influence the generation of cloud condensation nuclei, and thereby have an effect on the climate. The website ScienceDaily reported on a 2009 study that looked at how past changes in climate have been affected by the Earth's magnetic field. Geophysicist Mads Faurschou Knudsen, who co-authored the study, stated that the study's results supported Svensmark's theory. The authors of the study also acknowledged that CO_2 plays an important role in climate change.

Consensus View on Cosmic Rays

The view that cosmic rays could provide the mechanism by which changes in solar activity affect climate is not supported by the literature. Solomon *et al.* (2007) state:

[..] the cosmic ray time series does not appear to correspond to global total cloud cover after 1991 or to global low-level cloud cover after 1994. Together with the lack of a proven physical mechanism and the plausibility of other causal factors affecting changes in cloud cover, this makes the association between galactic cosmic ray-induced changes in aerosol and cloud formation controversial

Studies by Lockwood and Fröhlich (2007) and Sloan and Wolfendale (2008) found no relation between warming in recent decades and cosmic rays. Pierce and Adams (2009) used a model to simulate the effect of cosmic rays on cloud properties. They concluded that the hypothesized effect of cosmic rays was too small to explain recent climate change. Pierce and Adams (2009) noted that their findings did not rule out a possible connection between cosmic rays and climate change, and recommended further research.

Erlykin *et al.* (2009) found that the evidence showed that connections between solar variation and climate were more likely to be mediated by direct variation of insolation rather than cosmic rays, and concluded: "Hence within our assumptions, the effect of varying solar activity, either by direct solar irradiance or by varying cosmic ray rates, must be less than 0.07 °C since 1956, i.e. less than 14% of the observed global warming." Carslaw (2009) and Pittock (2009) review the recent and historical literature in this field and continue to find that the link between cosmic rays and climate is tenuous, though they encourage continued research. US EPA (2009) commented on research by Duplissy *et al.* (2009):

The CLOUD experiments at CERN are interesting research but do not provide conclusive evidence that cosmic rays can serve as a major source of cloud seeding. Preliminary results from the experiment (Duplissy et al., 2009) suggest that though there was some evidence of ion mediated nucleation, for most of the nucleation events observed the contribution of ion processes appeared to be minor. These experiments also showed the difficulty in maintaining sufficiently clean conditions and stable temperatures to prevent spurious aerosol bursts. There is no indication that the earlier Svensmark experiments could even have matched the controlled conditions of the CERN experiment. We find that the Svensmark results on cloud seeding have not yet been shown to be robust or sufficient to materially alter the conclusions of the assessment literature, especially given the abundance of recent literature that is skeptical of the cosmic ray-climate linkage

Effects of Global Warming

The effects of global warming are the environmental and social changes caused (directly or indirectly) by human emissions of greenhouse gases. There is a scientific consensus that climate change is occurring, and that human activities are the primary driver. Many impacts of climate change have already been observed, including glacier retreat, changes in the timing of seasonal events (e.g., earlier flowering of plants), and changes in agricultural productivity.

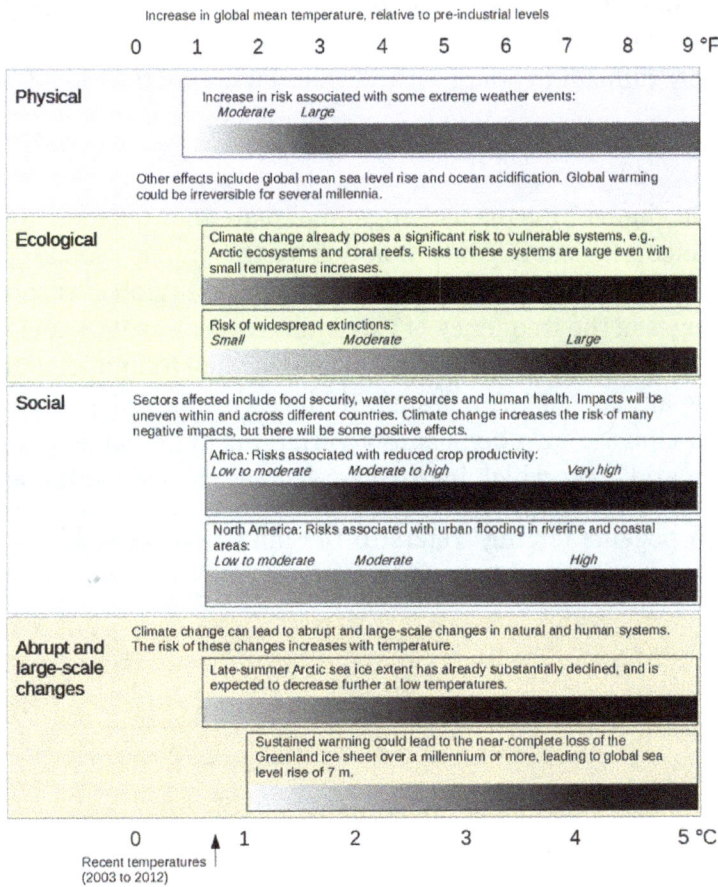

Summary of climate change impacts.

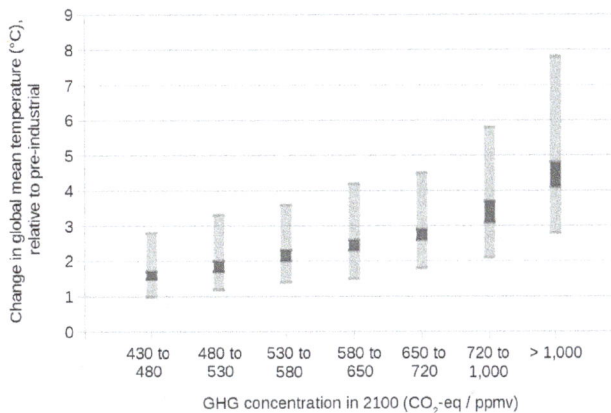

Projected global warming in 2100 for a range of emission scenarios.

Future effects of climate change will vary depending on climate change policies and social development. The two main policies to address climate change are reducing human greenhouse gas emissions (climate change mitigation) and adapting to the impacts of climate change. Geoengineering is another policy option.

Near-term climate change policies could significantly affect long-term climate change impacts. Stringent mitigation policies might be able to limit global warming (in 2100) to around 2 °C or below, relative to pre-industrial levels. Without mitigation, increased energy demand and extensive use of fossil fuels might lead to global warming of around 4 °C. Higher magnitudes of global warming would be more difficult to adapt to, and would increase the risk of negative impacts.

Definitions

In this article, "climate change" means a change in climate that persists over a sustained period of time. The World Meteorological Organization defines this time period as 30 years. Examples of climate change include increases in global surface temperature (global warming), changes in rainfall patterns, and changes in the frequency of extreme weather events. Changes in climate may be due to natural causes, e.g., changes in the sun's output, or due to human activities, e.g., changing the composition of the atmosphere. Any human-induced changes in climate will occur against a background of natural climatic variations and of variations in human activity such as population growth on shores or in arid areas which increase or decrease climate vulnerability.

Also, the term "anthropogenic forcing" refers to the influence exerted on a habitat or chemical environment by humans, as opposed to a natural process.

Temperature Changes

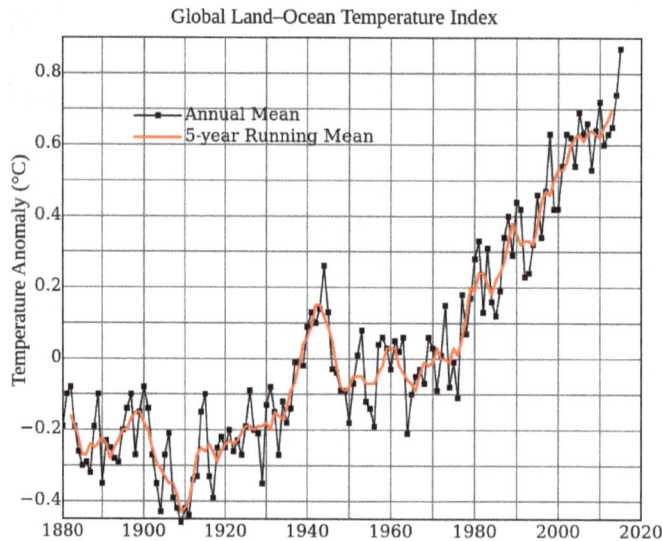

Global mean surface temperature change since 1880, relative to the 1951–1980 mean. Source: NASA GISS

The graph above shows the average of a set of temperature simulations for the 20th century (black line), followed by projected temperatures for the 21st century based on three greenhouse gas emissions scenarios (colored lines).

This article breaks down some of the impacts of climate change according to different levels of future global warming. This way of describing impacts has, for instance, been used in the IPCC (Intergovernmental Panel on Climate Change) Assessment Reports on climate change. The instrumental temperature record shows global warming of around 0.6 °C during the 20th century.

SRES Emissions Scenarios

The future level of global warming is uncertain, but a wide range of estimates (projections) have been made. The IPCC's "SRES" scenarios have been frequently used to make projections of future climate change. The SRES scenarios are "baseline" (or "reference") scenarios, which means that they do not take into account any current or future measures to limit GHG emissions (e.g., the UN-FCCC's Kyoto Protocol and the Cancún agreements). Emissions projections of the SRES scenarios are broadly comparable in range to the baseline emissions scenarios that have been developed by the scientific community.

In the IPCC Fourth Assessment Report, changes in future global mean temperature were projected using the six SRES "marker" emissions scenarios. Emissions projections for the six SRES "marker" scenarios are representative of the full set of forty SRES scenarios. For the lowest emissions SRES marker scenario, the best estimate for global mean temperature is an increase of 1.8 °C (3.2 °F) by the end of the 21st century. This projection is relative to global temperatures at the end of the 20th century. The "likely" range (greater than 66% probability, based on expert judgement) for the SRES B1 marker scenario is 1.1–2.9 °C (2.0–5.2 °F). For the highest emissions SRES marker scenario (A1FI), the best estimate for global mean temperature increase is 4.0 °C (7.2 °F), with a "likely" range of 2.4–6.4 °C (4.3–11.5 °F).

The range in temperature projections partly reflects (1) the choice of emissions scenario, and (2) the "climate sensitivity". For (1), different scenarios make different assumptions of future social and economic development (e.g., economic growth, population level, energy policies), which in turn affects projections of greenhouse gas (GHG) emissions. The projected magnitude of warming by 2100 is closely related to the level of cumulative emissions over the 21st century (i.e. total emissions between 2000-2100). The higher the cumulative emissions over this time period, the greater the level of warming is projected to occur.

(2) reflects uncertainty in the response of the climate system to past and future GHG emissions, which is measured by the climate sensitivity). Higher estimates of climate sensitivity lead to greater projected warming, while lower estimates of climate sensitivity lead to less projected warming.

Over the next several millennia, projections suggest that global warming could be irreversible. Even if emissions were drastically reduced, global temperatures would remain close to their highest level for at least 1,000 years.

Projected Warming in Context

Scientists have used various "proxy" data to assess past changes in Earth's climate (paleoclimate). Sources of proxy data include historical records (such as farmers' logs), tree rings, corals, fossil pollen, ice cores, and ocean and lake sediments. Analysis of these data suggest that recent warming is unusual in the past 400 years, possibly longer. By the end of the 21st century, temperatures may increase to a level not experienced since the mid-Pliocene, around 3 million years ago. At that time, models suggest that mean global temperatures were about 2–3 °C warmer than pre-industrial temperatures. Even a 2 °C rise above the pre-industrial level would be outside the range of temperatures experienced by human civilization.

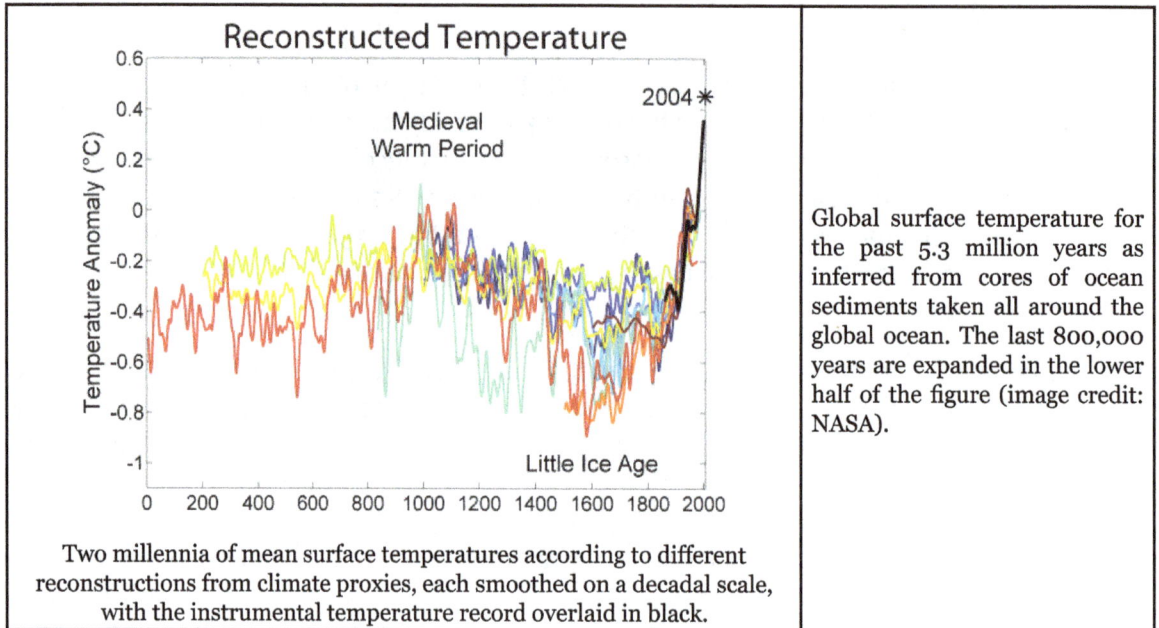

Reconstructed Temperature

Global surface temperature for the past 5.3 million years as inferred from cores of ocean sediments taken all around the global ocean. The last 800,000 years are expanded in the lower half of the figure (image credit: NASA).

Two millennia of mean surface temperatures according to different reconstructions from climate proxies, each smoothed on a decadal scale, with the instrumental temperature record overlaid in black.

Physical Impacts

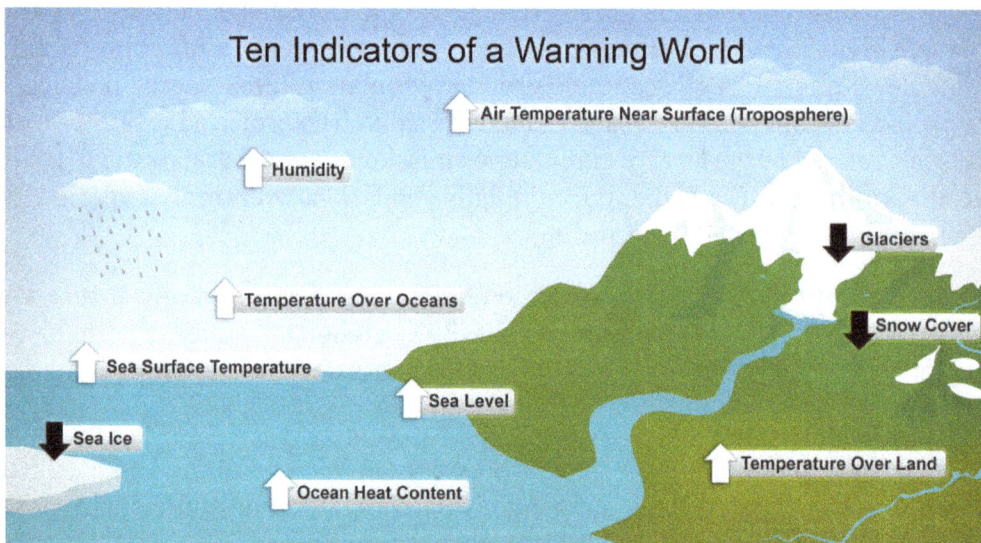

Seven of these indicators would be expected to increase in a warming world and observations show that they are, in fact, increasing. Three would be expected to decrease and they are, in fact, decreasing.

A broad range of evidence shows that the climate system has warmed. Evidence of global warming is shown in the graphs opposite. Some of the graphs show a positive trend, e.g., increasing temperature over land and the ocean, and sea level rise. Other graphs show a negative trend, e.g., decreased snow cover in the Northern Hemisphere, and declining Arctic sea ice extent. Evidence of warming is also apparent in living (biological) systems.

Human activities have contributed to a number of the observed changes in climate. This contribution has principally been through the burning of fossil fuels, which has led to an increase in the concentration of GHGs in the atmosphere. Another human influence on the climate are sulfur dioxide emissions, which are a precursor to the formation of sulfate aerosols in the atmosphere.

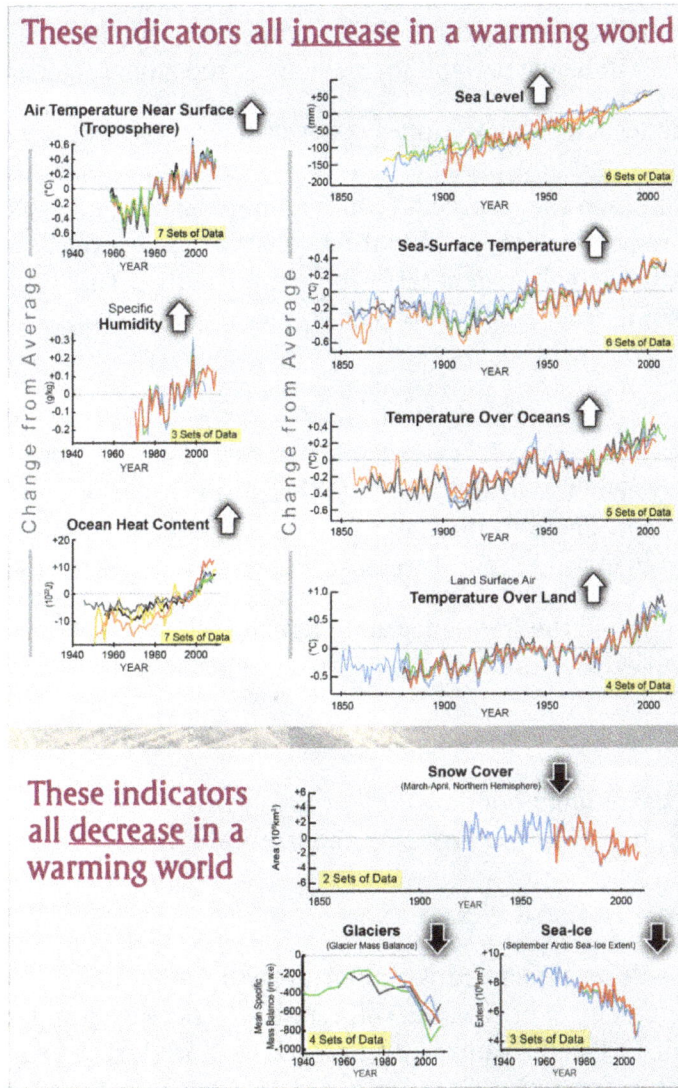

This set of graphs show changes in climate indicators over several decades. Each of the different colored lines in each panel represents an independently analyzed set of data. The data come from many different technologies including weather stations, satellites, weather balloons, ships and buoys.

Human-induced warming could lead to large-scale, irreversible, and/or abrupt changes in physical systems. An example of this is the melting of ice sheets, which contributes to sea level rise. The probability of warming having unforeseen consequences increases with the rate, magnitude, and duration of climate change.

Effects on Weather

Observations show that there have been changes in weather. As climate changes, the probabilities of certain types of weather events are affected.

Projected change in annual average precipitation by the end of the 21st century, based on a medium emissions scenario (SRES A1B).

Changes have been observed in the amount, intensity, frequency, and type of precipitation. Wide-spread increases in heavy precipitation have occurred, even in places where total rain amounts have decreased. With medium confidence, IPCC (2012) concluded that human influences had contributed to an increase in heavy precipitation events at the global scale.

Projections of future changes in precipitation show overall increases in the global average, but with substantial shifts in where and how precipitation falls. Projections suggest a reduction in rainfall in the subtropics, and an increase in precipitation in subpolar latitudes and some equatorial regions. In other words, regions which are dry at present will in general become even drier, while regions that are currently wet will in general become even wetter. This projection does not apply to every locale, and in some cases can be modified by local conditions.

Extreme Weather

Over most land areas since the 1950s, it is very likely that there have been fewer or warmer cold days and nights. Hot days and nights have also very likely become warmer or more frequent. Human activities have very likely contributed to these trends. There may have been changes in other climate extremes (e.g., floods, droughts and tropical cyclones) but these changes are more difficult to identify.

Projections suggest changes in the frequency and intensity of some extreme weather events. Confidence in projections varies over time.

Near-term Projections (2016–2035)

Some changes (e.g., more frequent hot days) will probably be evident in the near term, while other near-term changes (e.g., more intense droughts and tropical cyclones) are more uncertain.

Long-term Projections (2081–2100)

Future climate change will be associated with more very hot days and fewer very cold days. The frequency, length and intensity of heat waves will very likely increase over most land areas. Higher growth in anthropogenic GHG emissions will be associated with larger increases in the frequency and severity of temperature extremes.

Assuming high growth in GHG emissions (IPCC scenario RCP8.5), presently dry regions may be affected by an increase in the risk of drought and reductions in soil moisture. Over most of the mid-latitude land masses and wet tropical regions, extreme precipitation events will very likely become more intense and frequent.

Tropical Cyclones

At the global scale, the frequency of tropical cyclones will probably decrease or be unchanged. Global mean tropical cyclone maximum wind speed and precipitation rates will likely increase. Changes in tropical cyclones will probably vary by region, but these variations are uncertain.

Effects of Climate Extremes

The impacts of extreme events on the environment and human society will vary. Some impacts will be beneficial—e.g., fewer cold extremes will probably lead to fewer cold deaths. Overall, however, impacts will probably be mostly negative.

Cryosphere

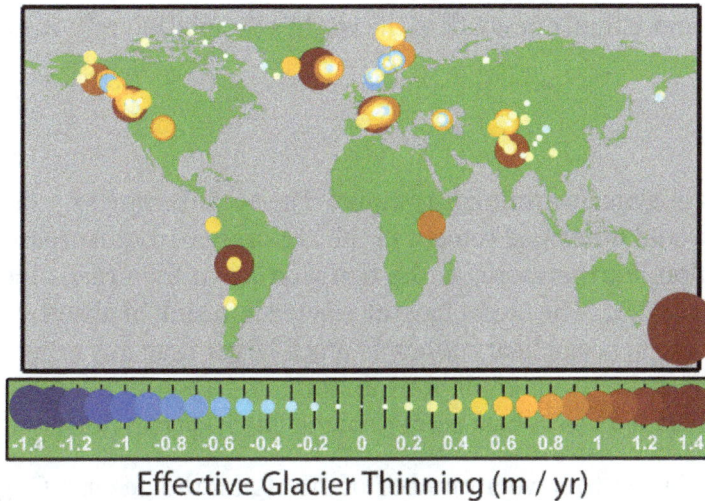

Mountain Glacier Changes Since 1970

Effective Glacier Thinning (m / yr)

A map of the change in thickness of mountain glaciers since 1970. Thinning in orange and red, thickening in blue.

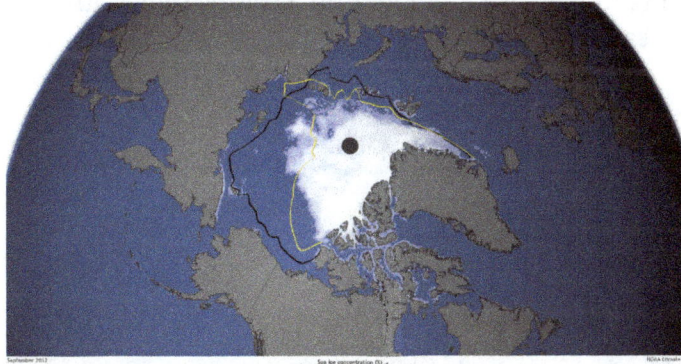

A map that shows ice concentration on 16 September 2012, along with the extent of the previous record low (yellow line) and the mid-September median extent (black line) setting a new record low that was 18 percent smaller than the previous record and nearly 50 percent smaller than the long-term (1979-2000) average.

The cryosphere is made up of areas of the Earth which are covered by snow or ice. Observed changes in the cryosphere include declines in Arctic sea ice extent, the widespread retreat of alpine glaciers, and reduced snow cover in the Northern Hemisphere.

Solomon *et al.* (2007) assessed the potential impacts of climate change on summertime Arctic sea ice extent. Assuming high growth in greenhouse gas emissions (SRES A2), some models projected that Arctic sea ice in the summer could largely disappear by the end of the 21st century. More recent projections suggest that the Arctic summers could be ice-free (defined as ice extent less than 1 million square km) as early as 2025-2030.

During the 21st century, glaciers and snow cover are projected to continue their widespread retreat. In the western mountains of North America, increasing temperatures and changes in precipitation are projected to lead to reduced snowpack. Snowpack is the seasonal accumulation of slow-melting snow. The melting of the Greenland and West Antarctic ice sheets could contribute to sea level rise, especially over long time-scales.

Changes in the cryosphere are projected to have social impacts. For example, in some regions, glacier retreat could increase the risk of reductions in seasonal water availability. Barnett *et al.* (2005) estimated that more than one-sixth of the world's population rely on glaciers and snowpack for their water supply.

Oceans

The role of the oceans in global warming is complex. The oceans serve as a sink for carbon dioxide, taking up much that would otherwise remain in the atmosphere, but increased levels of CO2 have led to ocean acidification. Furthermore, as the temperature of the oceans increases, they become less able to absorb excess CO2. The ocean have also acted as a sink in absorbing extra heat from the atmosphere. The increase in ocean heat content is much larger than any other store of energy in the Earth's heat balance over the two periods 1961 to 2003 and 1993 to 2003, and accounts for more than 90% of the possible increase in heat content of the Earth system during these periods.

Global warming is projected to have a number of effects on the oceans. Ongoing effects include rising sea levels due to thermal expansion and melting of glaciers and ice sheets, and warming of the ocean surface, leading to increased temperature stratification. Other possible effects include large-scale changes in ocean circulation.

Acidification

Changes in Aragonite Saturation of the World's Oceans, 1880–2012

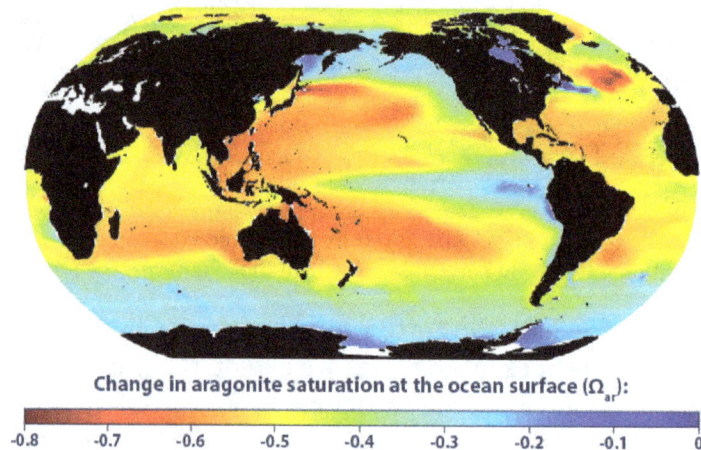

Change in aragonite saturation at the ocean surface (Ω_{ar}):

| -0.8 | -0.7 | -0.6 | -0.5 | -0.4 | -0.3 | -0.2 | -0.1 | 0 |

Data source: Feely, R.A., S.C. Doney, and S.R. Cooley. 2009. Ocean acidification: Present conditions and future changes in a high-CO_2 world. Oceanography 22(4):36–47.

For more information, visit U.S. EPA's "Climate Change Indicators in the United States" at www.epa.gov/climatechange/indicators.

This map shows changes in the amount of aragonite dissolved in ocean surface waters between the 1880s and the most recent decade (2003-2012). Historical modeling suggests that since the 1880s, increased CO_2 has led to lower aragonite saturation levels (less availability of minerals) in the oceans around the world. The largest decreases in aragonite saturation have occurred in tropical waters. However, decreases in cold areas may be of greater concern because colder waters typically have lower aragonite levels to begin with.

About one-third of the carbon dioxide emitted by human activity has already been taken up by the oceans. As carbon dioxide dissolves in sea water, carbonic acid is formed, which has the effect of acidifying the ocean, measured as a change in pH. The uptake of human carbon emissions since the year 1750 has led to an average decrease in pH of 0.1 units. Projections using the SRES emissions scenarios suggest a further reduction in average global surface ocean pH of between and 0.35 units over the 21st century.

The effects of ocean acidification on the marine biosphere have yet to be documented. Laboratory experiments suggest beneficial effects for a few species, with potentially highly detrimental effects for a substantial number of species. With medium confidence, Fischlin *et al.* (2007) projected that future ocean acidification and climate change would impair a wide range of planktonic and shallow benthic marine organisms that use aragonite to make their shells or skeletons, such as corals and marine snails (pteropods), with significant impacts particularly in the Southern Ocean.

Oxygen Depletion

The amount of oxygen dissolved in the oceans may decline, with adverse consequences for ocean life.

Sea Level Rise

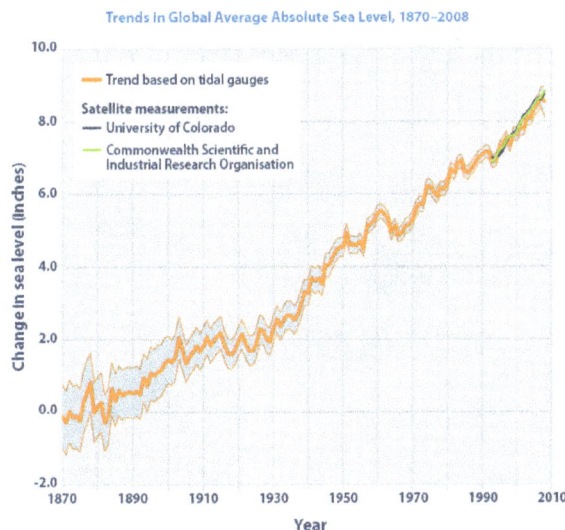

Trends in global average absolute sea level, 1870-2008.

There is strong evidence that global sea level rose gradually over the 20th century. With high confidence, Bindoff *et al.* (2007) concluded that between the mid-19th and mid-20th centuries, the rate of sea level rise increased. Authors of the IPCC Fourth Assessment Synthesis Report (IPCC AR4 SYR, 2007) reported that between the years 1961 and 2003, global average sea level rose at an average rate of 1.8 mm per year (mm/yr), with a range of 1.3–2.3 mm/yr. Between 1993 and 2003, the rate increased above the previous period to 3.1 mm/yr (range of 2.4–3.8 mm/yr). Authors of IPCC AR4 SYR (2007) were uncertain whether the increase in rate from 1993 to 2003 was due to natural variations in sea level over the time period, or whether it reflected an increase in the underlying long-term trend.

There are two main factors that have contributed to observed sea level rise. The first is thermal expansion: as ocean water warms, it expands. The second is from the contribution of land-based ice due to increased melting. The major store of water on land is found in glaciers and ice sheets. Anthropogenic forces very likely (greater than 90% probability, based on expert judgement) contributed to sea level rise during the latter half of the 20th century.

There is a widespread consensus that substantial long-term sea level rise will continue for centuries to come. In their Fourth Assessment Report, the IPCC projected sea level rise to the end of the 21st century using the SRES emissions scenarios. Across the six SRES marker scenarios, sea level was projected to rise by 18 to 59 cm (7.1 to 23.2 in), relative to sea level at the end of the 20th century. Thermal expansion is the largest component in these projections, contributing 70-75% of the central estimate for all scenarios. Due to a lack of scientific understanding, this sea level rise estimate does not include all of the possible contributions of ice sheets.

An assessment of the scientific literature on climate change was published in 2010 by the US National Research Council (US NRC, 2010). NRC (2010) described the projections in AR4 (i.e. those cited in the above paragraph) as "conservative", and summarized the results of more recent studies. Cited studies suggested a great deal of uncertainty in projections. A range of projections suggested possible sea level rise by the end of the 21st century of between 0.56 and 2 m, relative to sea levels at the end of the 20th century.

Ocean Temperature Rise

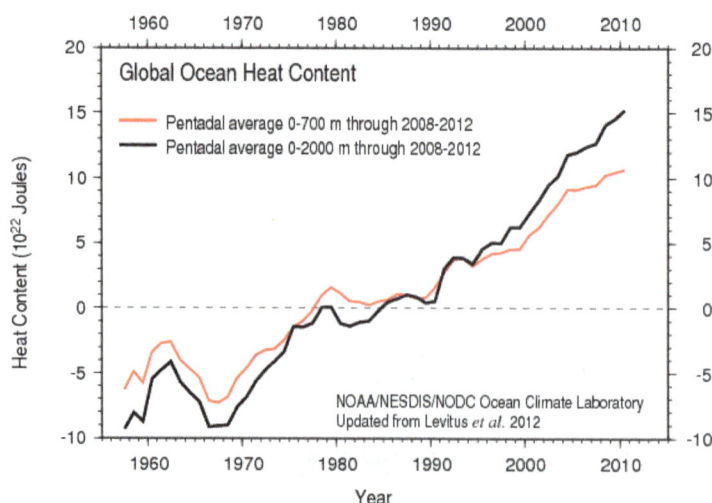

Global ocean heat content from 1955-2012

From 1961 to 2003, the global ocean temperature has risen by 0.10 °C from the surface to a depth of 700 m. There is variability both year-to-year and over longer time scales, with global ocean heat content observations showing high rates of warming for 1991–2003, but some cooling from 2003 to 2007. The temperature of the Antarctic Southern Ocean rose by 0.17 °C (0.31 °F) between the 1950s and the 1980s, nearly twice the rate for the world's oceans as a whole. As well as having effects on ecosystems (e.g. by melting sea ice, affecting algae that grow on its underside), warming reduces the ocean's ability to absorb CO2. It is likely (greater than 66% probability, based on expert judgement) that anthropogenic forcing contributed to the general warming observed in the upper several hundred metres of the ocean during the latter half of the 20th century.

Regions

Temperatures across the world in the 1880s (left) and the 1980s (right), as compared to average temperatures from 1951 to 1980.

Projected changes in average temperatures across the world in the 2050s under three greenhouse gas (GHG) emissions scenarios.

Regional effects of global warming vary in nature. Some are the result of a generalised global change, such as rising temperature, resulting in local effects, such as melting ice. In other cases, a change may be related to a change in a particular ocean current or weather system. In such cases, the regional effect may be disproportionate and will not necessarily follow the global trend.

There are three major ways in which global warming will make changes to regional climate: melting or forming ice, changing the hydrological cycle (of evaporation and precipitation) and changing currents in the oceans and air flows in the atmosphere. The coast can also be considered a region, and will suffer severe impacts from sea level rise.

Observed Impacts

With very high confidence, Rosenzweig *et al.* (2007) concluded that physical and biological systems on all continents and in most oceans had been affected by recent climate changes, particularly regional temperature increases. Impacts include earlier leafing of trees and plants over many regions; movements of species to higher latitudes and altitudes in the Northern Hemisphere; changes in bird migrations in Europe, North America and Australia; and shifting of the oceans' plankton and fish from cold- to warm-adapted communities.

The human influence on the climate can be seen in the geographical pattern of observed warming, with greater temperature increases over land and in polar regions rather than over the oceans. Using models, it is possible to identify the human "signal" of global warming over both land and ocean areas.

Projected Impacts

Projections of future climate changes at the regional scale do not hold as high a level of scientific confidence as projections made at the global scale. It is, however, expected that future warming will follow a similar geographical pattern to that seen already, with greatest warming over land and high northern latitudes, and least over the Southern Ocean and parts of the North Atlantic Ocean. Nearly all land areas will very likely warm more than the global average.

The Arctic, Africa, small islands and Asian megadeltas are regions that are likely to be especially affected by climate change. Low-latitude, less-developed areas are at most risk of experiencing negative impacts due to climate change. Developed countries are also vulnerable to climate change. For example, developed countries will be negatively affected by increases in the severity and frequency of some extreme weather events, such as heat waves. In all regions, some people can be particularly at risk from climate change, such as the poor, young children and the elderly.

Social Systems

The impacts of climate change can be thought of in terms of sensitivity and vulnerability. "Sensitivity" is the degree to which a particular system or sector might be affected, positively or negatively, by climate change and/or climate variability. "Vulnerability" is the degree to which a particular system or sector might be adversely affected by climate change.

The sensitivity of human society to climate change varies. Sectors sensitive to climate change include water resources, coastal zones, human settlements, and human health. Industries sensitive to climate change include agriculture, fisheries, forestry, energy, construction, insurance, financial services, tourism, and recreation.

Food Supply

Climate change will impact agriculture and food production around the world due to: the effects of elevated CO_2 in the atmosphere, higher temperatures, altered precipitation and transpiration regimes, increased frequency of extreme events, and modified weed, pest, and pathogen pressure. In general, low-latitude areas are at most risk of having decreased crop yields.

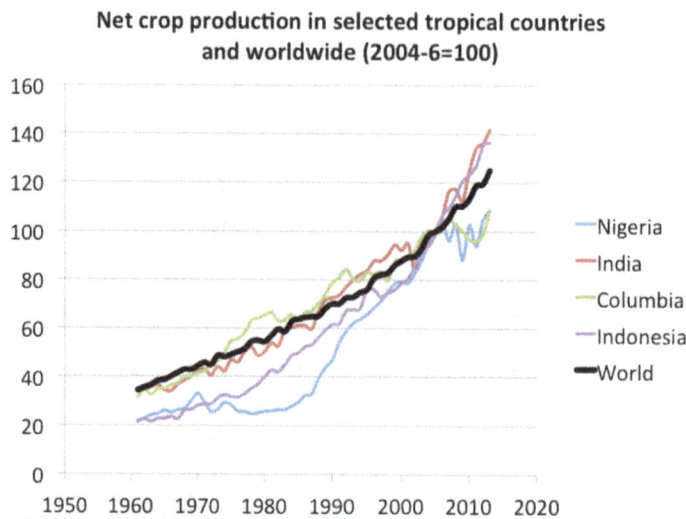

Graph of net crop production worldwide and in selected tropical countries. Raw data from the United Nations.

As of 2007, the effects of regional climate change on agriculture have been small. Changes in crop phenology provide important evidence of the response to recent regional climate change. Phenology is the study of natural phenomena that recur periodically, and how these phenomena relate to climate and seasonal changes. A significant advance in phenology has been observed for agriculture and forestry in large parts of the Northern Hemisphere.

Projections

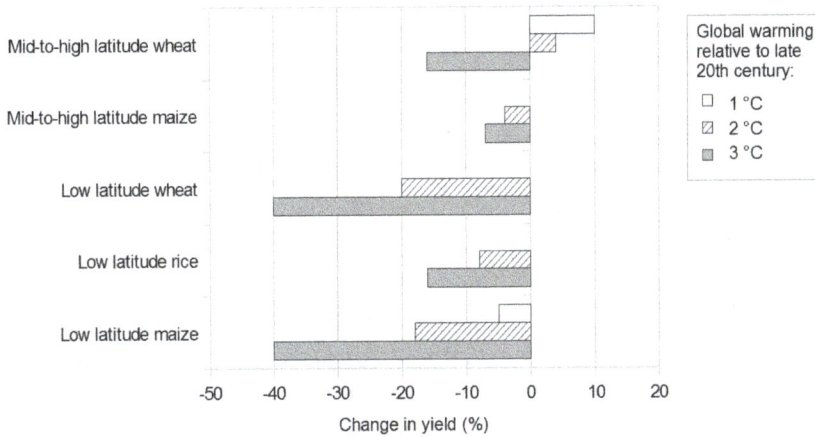

Projected changes in crop yields at different latitudes with global warming. This graph is based on several studies.

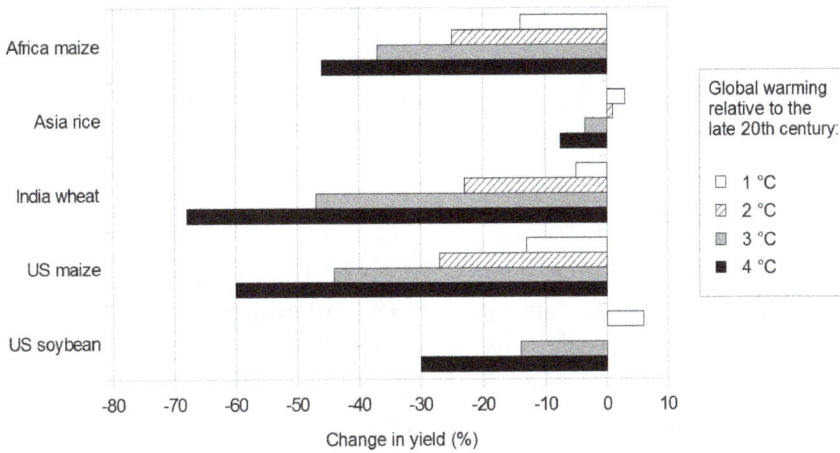

Projected changes in yields of selected crops with global warming. This graph is based on several studies.

With low to medium confidence, Schneider *et al.* (2007) projected that for about a 1 to 3 °C increase in global mean temperature (by the years 2090-2100, relative to average temperatures in the years 1990–2000), there would be productivity decreases for some cereals in low latitudes, and productivity increases in high latitudes. With medium confidence, global production potential was projected to:

- increase up to around 3 °C,

- very likely decrease above about 3 °C.

Most of the studies on global agriculture assessed by Schneider *et al.* (2007) had not incorporated a number of critical factors, including changes in extreme events, or the spread of pests and diseases. Studies had also not considered the development of specific practices or technologies to aid adaptation to climate change.

The graphs opposite show the projected effects of climate change on selected crop yields. Actual changes in yields may be above or below these central estimates.

The projections above can be expressed relative to pre-industrial (1750) temperatures. 0.6 °C of warming is estimated to have occurred between 1750 and 1990-2000. Add 0.6 °C to the above projections to convert them from a 1990-2000 to pre-industrial baseline.

Food Security

Easterling *et al.* (2007) assessed studies that made quantitative projections of climate change impacts on food security. It was noted that these projections were highly uncertain and had limitations. However, the assessed studies suggested a number of fairly robust findings. The first was that climate change would likely increase the number of people at risk of hunger compared with reference scenarios with no climate change. Climate change impacts depended strongly on projected future social and economic development. Additionally, the magnitude of climate change impacts was projected to be smaller compared to the impact of social and economic development. In 2006, the global estimate for the number of people undernourished was 820 million. Under the SRES A1, B1, and B2 scenarios, projections for the year 2080 showed a reduction in the number of people undernourished of about 560-700 million people, with a global total of undernourished people of 100-240 million in 2080. By contrast, the SRES A2 scenario showed only a small decrease in the risk of hunger from 2006 levels. The smaller reduction under A2 was attributed to the higher projected future population level in this scenario.

Droughts and Agriculture

Some evidence suggests that droughts have been occurring more frequently because of global warming and they are expected to become more frequent and intense in Africa, southern Europe, the Middle East, most of the Americas, Australia, and Southeast Asia. However, other research suggests that there has been little change in drought over the past 60 years. Their impacts are aggravated because of increased water demand, population growth, urban expansion, and environmental protection efforts in many areas. Droughts result in crop failures and the loss of pasture grazing land for livestock.

Health

Human beings are exposed to climate change through changing weather patterns (temperature, precipitation, sea-level rise and more frequent extreme events) and indirectly through changes in water, air and food quality and changes in ecosystems, agriculture, industry and settlements and the economy. According to an assessment of the scientific literature by Confalonieri *et al.* (2007), the effects of climate change to date have been small, but are projected to progressively increase in all countries and regions.

A study by the World Health Organization (WHO, 2009) estimated the effect of climate change on human health. Not all of the effects of climate change were included in their estimates, for example, the effects of more frequent and extreme storms were excluded. Climate change was estimated to have been responsible for 3% of diarrhoea, 3% of malaria, and 3.8% of dengue fever deaths worldwide in 2004. Total attributable mortality was about 0.2% of deaths in 2004; of these, 85% were child deaths.

Projections

With high confidence, authors of the IPCC AR4 Synthesis report projected that climate change would bring some benefits in temperate areas, such as fewer deaths from cold exposure, and some mixed effects such as changes in range and transmission potential of malaria in Africa. Benefits were projected to be outweighed by negative health effects of rising temperatures, especially in developing countries.

With very high confidence, Confalonieri *et al.* (2007) concluded that economic development was an important component of possible adaptation to climate change. Economic growth on its own, however, was not judged to be sufficient to insulate the world's population from disease and injury due to climate change. Future vulnerability to climate change will depend not only on the extent of social and economic change, but also on how the benefits and costs of change are distributed in society. For example, in the 19th century, rapid urbanization in western Europe lead to a plummeting in population health. Other factors important in determining the health of populations include education, the availability of health services, and public-health infrastructure.

Water Resources

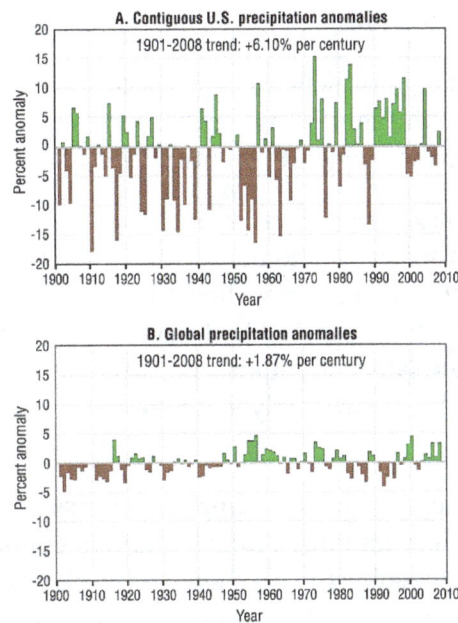

a Anomalies and percent change are calculated with respect to the 1971-2000 mean.
Data source: NOAA, 2009

Precipitation during the 20th century and up through 2008 during global warming, the NOAA estimating an observed trend over that period of 1.87% global precipitation increase per century.

A number of climate-related trends have been observed that affect water resources. These include changes in precipitation, the crysosphere and surface waters (e.g., changes in river flows). Observed and projected impacts of climate change on freshwater systems and their management are mainly due to changes in temperature, sea level and precipitation variability. Sea level rise will extend areas of salinization of groundwater and estuaries, resulting in a decrease in freshwater availability for humans and ecosystems in coastal areas. In an assessment of the scientific literature, Kundzewicz *et al.* (2007) concluded, with high confidence, that:

- the negative impacts of climate change on freshwater systems outweigh the benefits. All of the regions assessed in the IPCC Fourth Assessment Report (Africa, Asia, Australia and New Zealand, Europe, Latin America, North America, Polar regions (Arctic and Antarctic), and small islands) showed an overall net negative impact of climate change on water resources and freshwater ecosystems.

- Semi-arid and arid areas are particularly exposed to the impacts of climate change on freshwater. With very high confidence, it was judged that many of these areas, e.g., the Mediterranean basin, Western United States, Southern Africa, and north-eastern Brazil, would suffer a decrease in water resources due to climate change.

Migration and Conflict

General circulation models project that the future climate change will bring wetter coasts, drier mid-continent areas, and further sea level rise. Such changes could result in the gravest effects of climate change through human migration. Millions might be displaced by shoreline erosions, river and coastal flooding, or severe drought.

Migration related to climate change is likely to be predominantly from rural areas in developing countries to towns and cities. In the short term climate stress is likely to add incrementally to existing migration patterns rather than generating entirely new flows of people.

It has been argued that environmental degradation, loss of access to resources (e.g., water resources), and resulting human migration could become a source of political and even military conflict. Factors other than climate change may, however, be more important in affecting conflict. For example, Wilbanks *et al.* (2007) suggested that major environmentally influenced conflicts in Africa were more to do with the relative abundance of resources, e.g., oil and diamonds, than with resource scarcity. Scott *et al.* (2001) placed only low confidence in predictions of increased conflict due to climate change.

A 2013 study found that significant climatic changes were associated with a higher risk of conflict worldwide, and predicted that "amplified rates of human conflict could represent a large and critical social impact of anthropogenic climate change in both low- and high-income countries." Similarly, a 2014 study found that higher temperatures were associated with a greater likelihood of violent crime, and predicted that global warming would cause millions of such crimes in the United States alone during the 21st century.

Military planners are concerned that global warming is a "threat multiplier". "Whether it is poverty, food and water scarcity, diseases, economic instability, or threat of natural disasters, the broad range of changing climatic conditions may be far reaching. These challenges may threaten stability in much of the world".

Aggregate Impacts

Aggregating impacts adds up the total impact of climate change across sectors and/or regions.

Examples of aggregate measures include economic cost (e.g., changes in gross domestic product (GDP) and the social cost of carbon), changes in ecosystems (e.g., changes over land area from one type of vegetation to another), human health impacts, and the number of people affected by climate change. Aggregate measures such as economic cost require researchers to make value judgements over the importance of impacts occurring in different regions and at different times.

Observed Impacts

Global losses reveal rapidly rising costs due to extreme weather-related events since the 1970s. Socio-economic factors have contributed to the observed trend of global losses, e.g., population growth, increased wealth. Part of the growth is also related to regional climatic factors, e.g., changes in precipitation and flooding events. It is difficult to quantify the relative impact of socio-economic factors and climate change on the observed trend. The trend does, however, suggest increasing vulnerability of social systems to climate change.

Projected Impacts

The total economic impacts from climate change are highly uncertain. With medium confidence, Smith *et al.* (2001) concluded that world GDP would change by plus or minus a few percent for a small increase in global mean temperature (up to around 2 °C relative to the 1990 temperature level). Most studies assessed by Smith *et al.* (2001) projected losses in world GDP for a medium increase in global mean temperature (above 2-3 °C relative to the 1990 temperature level), with increasing losses for greater temperature increases. This assessment is consistent with the findings of more recent studies, as reviewed by Hitz and Smith (2004).

Economic impacts are expected to vary regionally. For a medium increase in global mean temperature (2-3 °C of warming, relative to the average temperature between 1990–2000), market sectors in low-latitude and less-developed areas might experience net costs due to climate change. On the other hand, market sectors in high-latitude and developed regions might experience net benefits for this level of warming. A global mean temperature increase above about 2-3 °C (relative to 1990-2000) would very likely result in market sectors across all regions experiencing either declines in net benefits or rises in net costs.

Aggregate impacts have also been quantified in non-economic terms. For example, climate change over the 21st century is likely to adversely affect hundreds of millions of people through increased coastal flooding, reductions in water supplies, increased malnutrition and increased health impacts.

Biological Systems

Observed Impacts on Biological Systems

A vast array of physical and biological systems across the Earth are being affected by human-induced global warming.

With very high confidence, Rosenzweig *et al.* (2007) concluded that recent warming had strongly affected natural biological systems. Hundreds of studies have documented responses of ecosystems, plants, and animals to the climate changes that have already occurred. For example, in the

Northern Hemisphere, species are almost uniformly moving their ranges northward and up in elevation in search of cooler temperatures. Humans are very likely causing changes in regional temperatures to which plants and animals are responding.

Projected Impacts on Biological Systems

By the year 2100, ecosystems will be exposed to atmospheric CO_2 levels substantially higher than in the past 650,000 years, and global temperatures at least among the highest of those experienced in the past 740,000 years. Significant disruptions of ecosystems are projected to increase with future climate change. Examples of disruptions include disturbances such as fire, drought, pest infestation, invasion of species, storms, and coral bleaching events. The stresses caused by climate change, added to other stresses on ecological systems (e.g., land conversion, land degradation, harvesting, and pollution), threaten substantial damage to or complete loss of some unique eco-systems, and extinction of some critically endangered species.

Climate change has been estimated to be a major driver of biodiversity loss in cool conifer forests, savannas, mediterranean-climate systems, tropical forests, in the Arctic tundra, and in coral reefs. In other ecosystems, land-use change may be a stronger driver of biodiversity loss at least in the near-term. Beyond the year 2050, climate change may be the major driver for biodiversity loss globally.

A literature assessment by Fischlin *et al.* (2007) included a quantitative estimate of the number of species at increased risk of extinction due to climate change. With medium confidence, it was projected that approximately 20 to 30% of plant and animal species assessed so far (in an unbiased sample) would likely be at increasingly high risk of extinction should global mean temperatures exceed a warming of 2 to 3 °C above pre-industrial temperature levels. The uncertainties in this estimate, however, are large: for a rise of about 2 °C the percentage may be as low as 10%, or for about 3 °C, as high as 40%, and depending on biota (all living organisms of an area, the flora and fauna considered as a unit) the range is between 1% and 80%. As global average temperature exceeds 4 °C above pre-industrial levels, model projections suggested that there could be significant extinctions (40-70% of species that were assessed) around the globe.

Assessing whether future changes in ecosystems will be beneficial or detrimental is largely based on how ecosystems are valued by human society. For increases in global average temperature exceeding 1.5 to 2.5 °C (relative to global temperatures over the years 1980-1999) and in concomitant atmospheric CO_2 concentrations, projected changes in ecosystems will have predominantly negative consequences for biodiversity and ecosystems goods and services, e.g., water and food supply.

Abrupt or Irreversible Changes

Physical, ecological and social systems may respond in an abrupt, non-linear or irregular way to climate change. This is as opposed to a smooth or regular response. A quantitative entity behaves "irregularly" when its dynamics are discontinuous (i.e., not smooth), nondifferentiable, unbounded, wildly varying, or otherwise ill-defined. Such behaviour is often termed "singular". Irregular behaviour in Earth systems may give rise to certain thresholds, which, when crossed, may lead to a large change in the system.

Some singularities could potentially lead to severe impacts at regional or global scales. Examples of "large-scale" singularities are discussed in the articles on abrupt climate change, climate change feedback and runaway climate change. It is possible that human-induced climate change could trigger large-scale singularities, but the probabilities of triggering such events are, for the most part, poorly understood.

With low to medium confidence, Smith *et al.* (2001) concluded that a rapid warming of more than 3 °C above 1990 levels would exceed thresholds that would lead to large-scale discontinuities in the climate system. Since the assessment by Smith *et al.* (2001), improved scientific understanding provides more guidance for two large-scale singularities: the role of carbon cycle feedbacks in future climate change and the melting of the Greenland and West Antarctic ice sheets.

Biogeochemical Cycles

Climate change may have an effect on the carbon cycle in an interactive "feedback" process. A feedback exists where an initial process triggers changes in a second process that in turn influences the initial process. A positive feedback intensifies the original process, and a negative feedback reduces it. Models suggest that the interaction of the climate system and the carbon cycle is one where the feedback effect is positive.

Using the A2 SRES emissions scenario, Schneider *et al.* (2007) found that this effect led to additional warming by the years 2090-2100 (relative to the 1990–2000) of 0.1–1.5 °C. This estimate was made with high confidence. The climate projections made in the IPCC Fourth Assessment Report summarized earlier of 1.1–6.4 °C account for this feedback effect. On the other hand, with medium confidence, Schneider *et al.* (2007) commented that additional releases of GHGs were possible from permafrost, peat lands, wetlands, and large stores of marine hydrates at high latitudes.

Greenland and West Antarctic Ice Sheets

With medium confidence, authors of AR4 concluded that with a global average temperature increase of 1–4 °C (relative to temperatures over the years 1990–2000), at least a partial deglaciation of the Greenland ice sheet, and possibly the West Antarctic ice sheets would occur. The estimated timescale for partial deglaciation was centuries to millennia, and would contribute 4 to 6 metres (13 to 20 ft) or more to sea level rise over this period.

Atlantic Meridional Overturning Circulation

The Atlantic Meridional Overturning Circulation (AMOC) is an important component of the Earth's climate system, characterized by a northward flow of warm, salty water in the upper layers of the Atlantic and a southward flow of colder water in the deep Atlantic. The AMOC is equivalently known as the thermohaline circulation (THC). Potential impacts associated with MOC changes include reduced warming or (in the case of abrupt change) absolute cooling of northern high-latitude areas near Greenland and north-western Europe, an increased warming of Southern Hemisphere high-latitudes, tropical drying, as well as changes to marine ecosystems, terrestrial vegetation, oceanic CO2 uptake, oceanic oxygen concentrations, and shifts in fisheries. According to an assessment by the US Climate Change Science Program (CCSP, 2008b), it is very likely (greater than 90% probability, based on expert judgement) that the strength of the AMOC will decrease over the course of the 21st century. Warming is still expected to occur over most of the European region downstream of the North Atlantic Current in response to increasing GHGs, as well as over North America. Although it is very unlikely (less than 10% probability, based on expert judgement) that the AMOC will collapse in the 21st century, the potential consequences of such a collapse could be severe.

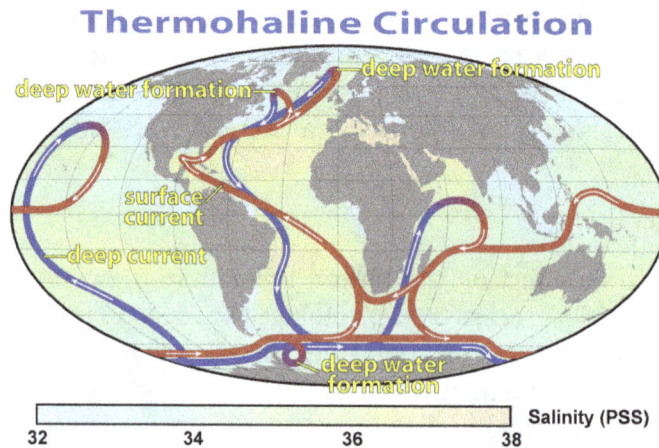

Thermohaline Circulation

This map shows the general location and direction of the warm surface (red) and cold deep water (blue) currents of the thermohaline circulation. Salinity is represented by color in units of the Practical Salinity Scale. Low values (blue) are less saline, while high values (orange) are more saline.

Irreversibilities

Commitment to Radiative Forcing

Emissions of GHGs are a potentially irreversible commitment to sustained radiative forcing in the future. The contribution of a GHG to radiative forcing depends on the gas's ability to trap infrared (heat) radiation, the concentration of the gas in the atmosphere, and the length of time the gas resides in the atmosphere.

CO_2 is the most important anthropogenic GHG. While more than half of the CO_2 emitted is currently removed from the atmosphere within a century, some fraction (about 20%) of emitted CO_2 remains in the atmosphere for many thousands of years. Consequently, CO_2 emitted today is potentially an irreversible commitment to sustained radiative forcing over thousands of years.

This commitment may not be truly irreversible should techniques be developed to remove CO_2 or other GHGs directly from the atmosphere, or to block sunlight to induce cooling. Techniques of this sort are referred to as geoengineering. Little is known about the effectiveness, costs or potential side-effects of geoengineering options. Some geoengineering options, such as blocking sunlight, would not prevent further ocean acidification.

Irreversible Impacts

Human-induced climate change may lead to irreversible impacts on physical, biological, and social systems. There are a number of examples of climate change impacts that may be irreversible, at least over the timescale of many human generations. These include the large-scale singularities described above – changes in carbon cycle feedbacks, the melting of the Greenland and West Antarctic ice sheets, and changes to the AMOC. In biological systems, the extinction of species would be an irreversible impact. In social systems, unique cultures may be lost due to climate change. For example, humans living on atoll islands face risks due to sea-level rise, sea-surface warming, and increased frequency and intensity of extreme weather events.

Benefits of Global Warming

With a large range of effects, it is unlikely that all effects will be negative. Not only have some positive effects been identified, but there is some published material indicating that a small amount of warming would be good. The IPCC cautions that "Estimates agree on the size of the impact (small relative to economic growth), and 17 of the 20 impact estimates shown in Figure 10-1 are negative. Losses accelerate with greater warming, and estimates diverge."

The identified benefits are listed below.

CO2 Fertilisation Effect

CO_2 is one of the substances which plants require to grow. Increasing its amount in the air contributes to:

- Improved agriculture in some high latitude regions
- Increased growing season in Greenland
- Increased productivity of sour orange trees
- Increased vegetation activity in high northern latitudes
- Increased plankton biomass in the North Pacific Subtropical Gyre
- Recent increase in forest growth
- Increased Arctic tundra plant reproduction

Human Health

- Winter deaths will decline as temperatures warm

Ice-free Northwest Passage

- Ships will travel on a shorter route between the Pacific and Atlantic oceans

Animal Population Changes

Some animals will benefit from the warming:

- Increase in chinstrap and gentoo penguins
- Bigger marmots

Scientific Opinion

The Intergovernmental Panel on Climate Change (IPCC) has published several major assessments on the effects of global warming. Its most recent comprehensive impact assessment was published in 2014. Publications describing the effects of climate change have also been produced by the following organizations:

- American Association for the Advancement of Science (AAAS)

- A report by the Netherlands Environmental Assessment Agency, the Royal Netherlands Meteorological Institute, and Wageningen University and Research Centre

- UK AVOID research programme

- A report by the UK Royal Society and US National Academy of Sciences

- University of New South Wales Climate Change Research Centre

- US National Research Council

A report by Molina *et al.* (no date) states:

The overwhelming evidence of human-caused climate change documents both current impacts with significant costs and extraordinary future risks to society and natural systems

Nasa Data and Tools

NASA has released public data and tools to predict how temperature and rainfall patterns worldwide may change through to the year 2100 caused by increasing carbon dioxide in Earth's atmosphere. The dataset shows projected changes worldwide on a regional level simulated by 21 climate models. The data can be viewed on a daily timescale for individual cities and towns and may be used to conduct climate risk assessments to predict the local and global effects of weather dangers, for example droughts, floods, heat waves and declines in agriculture productivity, and help plan responses to global warming effects.

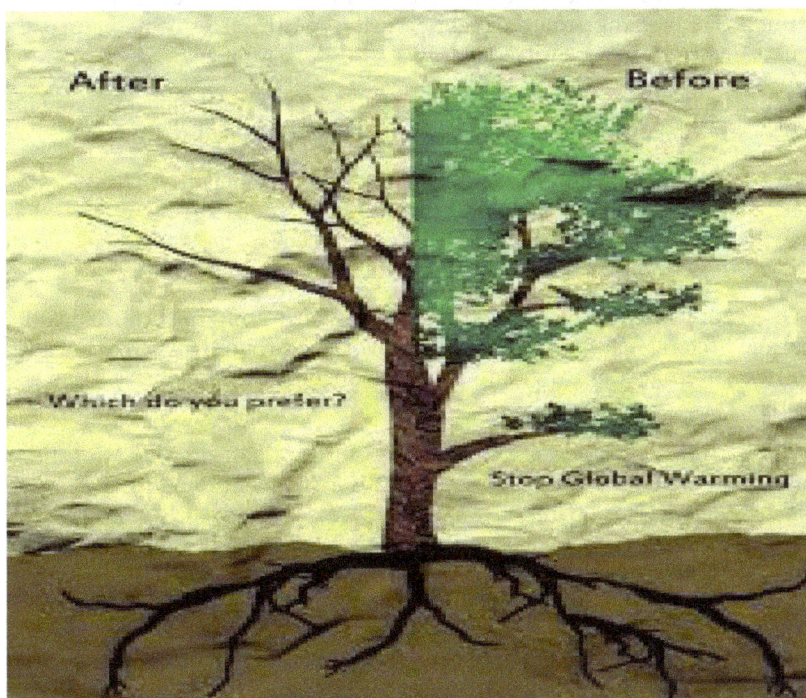

References

- Houghton, J. (2002), "An Overview of the Intergovernmental Panel on Climate Change (IPCC) and Its Process of Science Assessment", in Hester, R.E.; Harrison, R.M., Issues in Environmental Science and Technology (PDF), Global Environmental Change, 17, The Royal Society of Chemistry, ISBN 978-0-85404-280-7 .

- IPCC SAR WG1 (1996), Houghton, J.T.; Meira Filho, L.G.; Callander, B.A.; Harris, N.; Kattenberg, A.; Maskell, K., eds., Climate Change 1995: The Science of Climate Change (PDF), Contribution of Working Group I to the Second Assessment Report of the Intergovernmental Panel on Climate Change, Cambridge University Press, ISBN 0-521-56433-6 (pb: 0-521-56436-0).

- IPCC TAR WG1 (2001), Houghton, J.T.; Ding, Y.; Griggs, D.J.; Noguer, M.; van der Linden, P.J.; Dai, X.; Maskell, K.; Johnson, C.A., eds., Climate Change 2001: The Scientific Basis, Contribution of Working Group I to the Third Assessment Report of the Intergovernmental Panel on Climate Change, Cambridge University Press, ISBN 0-521-80767-0 (pb: 0-521-01495-6).

- IPCC AR4 WG1 (2007), Solomon, S.; Qin, D.; Manning, M.; Chen, Z.; Marquis, M.; Averyt, K.B.; Tignor, M.; Miller, H.L., eds., Climate Change 2007: The Physical Science Basis, Contribution of Working Group I to the Fourth Assessment Report of the Intergovernmental Panel on Climate Change, Cambridge University Press, ISBN 978-0-521-88009-1 (pb: 978-0-521-70596-7).

- IPCC SAR WG2 (1996), Watson, R.T.; Zinyowera, M.C.; Moss, R.H., eds., Climate Change 1995: Impacts, Adaptations and Mitigation of Climate Change: Scientific-Technical Analyses, Contribution of Working Group II to the Second Assessment Report of the Intergovernmental Panel on Climate Change, Cambridge University Press, ISBN 0-521-56431-X (pb: 0-521-56437-9 pdf.

- IPCC AR4 WG3 (2007), Metz, B.; Davidson, O.R.; Bosch, P.R.; Dave, R.; Meyer, L.A., eds., Climate Change 2007: Mitigation of Climate Change, Contribution of Working Group III to the Fourth Assessment Report of the Intergovernmental Panel on Climate Change, Cambridge University Press, ISBN 978-0-521-88011-4 (pb: 978-0-521-70598-1).

- IPCC AR4 SYR (2007), Core Writing Team; Pachauri, R.K; Reisinger, A., eds., Climate Change 2007: Synthesis Report, Contribution of Working Groups I, II and III to the Fourth Assessment Report of the Intergovernmental Panel on Climate Change, Geneva, Switzerland: IPCC, ISBN 92-9169-122-4 .

- Karl, Thomas R.; Melillo, Jerry M.; Peterson, Thomas C., eds. (2009). Global Climate Change Impacts in the United States (PDF). New York: Cambridge University Press. ISBN 978-0-521-14407-0.

- Staff of the International Bank for Reconstruction and Development / The World Bank (2010). World Development Report 2010: Development and Climate Change. 1818 H Street NW, Washington DC 20433, USA: International Bank for Reconstruction and Development / The World Bank. doi:10.1596/978-0-8213-7987-5. ISBN 978-0-8213-7987-5.

- IPCC AR5 WG2 A (2014), Field, C.B.; et al., eds., Climate Change 2014: Impacts, Adaptation, and Vulnerability. Part A: Global and Sectoral Aspects. Contribution of Working Group II (WG2) to the Fifth Assessment Report (AR5) of the Intergovernmental Panel on Climate Change (IPCC), Cambridge University Press . Archived 20 October 2014.

- IPCC AR5 WG3 (2014), Edenhofer, O.; et al., eds., Climate Change 2014: Mitigation of Climate Change. Contribution of Working Group III (WG3) to the Fifth Assessment Report (AR5) of the Intergovernmental Panel on Climate Change (IPCC), Cambridge University Press . Archived 29 June 2014.

Tool of Measuring Greenhouse Gas

Carbon accounting is a method for measuring greenhouse gas emissions. It has been taken up by corporations and firms that aim to reduce greenhouse gas emission as well as stimulate demand for carbon credit commodities. However, accurate measurement of greenhouse gases is still extremely difficult. This chapter discusses the methods of greenhouse gas measurement in a critical manner, providing key analysis to the subject matter.

Carbon Accounting

Carbon accounting refers generally to processes undertaken to "measure" amounts of carbon dioxide equivalents emitted by an entity. It is used inter alia by nation states, corporations, individuals – to create the carbon credit commodity traded on carbon markets (or to establish the demand for carbon credits). Correspondingly, examples for products based upon forms of carbon accounting can be found in national inventories, corporate environmental reports or carbon footprint calculators. Likening sustainability measurement, as an instance of ecological modernisation discourses and policy, carbon accounting is hoped to provide a factual ground for carbon-related decision-making. However, social scientific studies of accounting challenge this hope, pointing to the socially constructed character of carbon conversion factors or of the accountants' work practice which cannot implement abstract accounting schemes into reality.

While natural sciences claim to know and measure carbon, for organisations it is usually easier to employ forms of carbon accounting to represent carbon. The trustworthiness of accounts of carbon emissions can easily be contested. Thus, how well carbon accounting represents carbon is difficult to exactly know. Science and Technology Studies scholar Donna Haraway's pluralised concept of knowledge, i.e. knowledges, can well be used to understand better the status of knowledge produced by carbon accounting: carbon accounting produced a version of understanding of carbon emissions. Other carbon accountants would produce other results.

Carbon Accounting in Corporations

Carbon accounting can be used as part of sustainability accounting by for-profit and non-profit organisations. A corporate or organisational "carbon" or greenhouse gas (GHG) emissions assessment promises to quantify the greenhouse gases produced directly and indirectly from a business or organisation's activities within a set of boundaries. Also known as a carbon footprint, it is a business tool that constructs information that may (or may not) be useful for understanding and managing climate change impacts.

The drivers for corporate carbon accounting include mandatory GHG reporting in directors' reports, investment due diligence, shareholder and stakeholder communication, staff engagement, green messaging, and tender requirements for business and government contracts. Accounting for greenhouse gas emissions is increasingly framed as a standard requirement for business. As of June 2011, 60% of UK FTSE 100 companies had published environmental targets, with 53% of these 240+ targets relating to carbon, greenhouse gas emissions or energy reductions (representing 59% of the FTSE 100). In June 2012, the UK coalition government announced the introduction of mandatory carbon reporting, requiring around 1,100 of the UK's largest listed companies to report their greenhouse gas emissions every year. Deputy Prime Minister Nick Clegg confirmed that emission reporting rules would come into effect from April 2013 in his piece for The Guardian.

Carbon Accounting of Avoided Emissions

A special case of carbon accounting is the accounting process undertaken to measure the amount of carbon dioxide equivalents that will not be released into the atmosphere as a result of flexible mechanisms projects under the Kyoto Protocol. These projects thus include (but are not limited to) renewable energy projects and biomass, forage and tree plantations.

Carbon Accounting Software

A number of programs are created in order to assist with carbon accounting.

References

- I. Lippert. Enacting Environments: An Ethnography of the Digitalisation and Naturalisation of Emissions. University of Augsburg, 2013.

- I. Lippert. Extended carbon cognition as a machine. Computational Culture, 1(1), 2011. and I. Lippert. Carbon classified? Unpacking heterogeneous relations inscribed into corporate carbon emissions. Ephemera, 12(1/2):138–161, 2012.

- I. Lippert. Carbon dioxide. In C. A. Zimring, editor, Encyclopedia of Consumption and Waste: The Social Science of Garbage. Sage Publications, Feb. 2012.

- "Raising The Bar - Building sustainable business value through environmental targets". Carbon Trust. June 2011. Retrieved 2012-11-12.

- "Rio's reprise must set hard deadlines for development". The Guardian. 2012-06-19. Archived from the original on July 30, 2012. Retrieved 2012-07-30.

Permissions

All chapters in this book are published with permission under the Creative Commons Attribution Share Alike License or equivalent. Every chapter published in this book has been scrutinized by our experts. Their significance has been extensively debated. The topics covered herein carry significant information for a comprehensive understanding. They may even be implemented as practical applications or may be referred to as a beginning point for further studies.

We would like to thank the editorial team for lending their expertise to make the book truly unique. They have played a crucial role in the development of this book. Without their invaluable contributions this book wouldn't have been possible. They have made vital efforts to compile up to date information on the varied aspects of this subject to make this book a valuable addition to the collection of many professionals and students.

This book was conceptualized with the vision of imparting up-to-date and integrated information in this field. To ensure the same, a matchless editorial board was set up. Every individual on the board went through rigorous rounds of assessment to prove their worth. After which they invested a large part of their time researching and compiling the most relevant data for our readers.

The editorial board has been involved in producing this book since its inception. They have spent rigorous hours researching and exploring the diverse topics which have resulted in the successful publishing of this book. They have passed on their knowledge of decades through this book. To expedite this challenging task, the publisher supported the team at every step. A small team of assistant editors was also appointed to further simplify the editing procedure and attain best results for the readers.

Apart from the editorial board, the designing team has also invested a significant amount of their time in understanding the subject and creating the most relevant covers. They scrutinized every image to scout for the most suitable representation of the subject and create an appropriate cover for the book.

The publishing team has been an ardent support to the editorial, designing and production team. Their endless efforts to recruit the best for this project, has resulted in the accomplishment of this book. They are a veteran in the field of academics and their pool of knowledge is as vast as their experience in printing. Their expertise and guidance has proved useful at every step. Their uncompromising quality standards have made this book an exceptional effort. Their encouragement from time to time has been an inspiration for everyone.

The publisher and the editorial board hope that this book will prove to be a valuable piece of knowledge for students, practitioners and scholars across the globe.

Index

www.ingramcontent.com/pod-product-compliance
Lightning Source LLC
Chambersburg PA
CBHW082019190326
41458CB00010B/3230